Aspergillus and Aspergillosis - Advances in Genomics, Drug Development, Diagnosis and Treatment

Edited by Mehdi Razzaghi-Abyaneh, Mahendra Rai and Masoomeh Shams-Ghahfarokhi

Published in London, United Kingdom

Aspergillus and Aspergillosis - Advances in Genomics, Drug Development, Diagnosis and Treatment
http://dx.doi.org/10.5772/intechopen.107748
Edited by Mehdi Razzaghi-Abyaneh, Mahendra Rai and Masoomeh Shams-Ghahfarokhi

Contributors
Mohammadreza Salehi, Sadegh Khodavaisy, Fariba Zamani, Nur Azura Mohd Said, Noor Sheryna Jusoh, Faridah Salam, Lily Sohaida Mohd Sojak, Norhafniza Awaludin, Mohammad Rejab Ismail, Noor Fadhilah Mohd Bakri, Marcela Patricia Gomez Rojas, Óscar Marino Mosquera Martinez, Amir Afzal, Sairah Syed, Rafiq Ahmad, Muhammad Zeeshan, Ghulam Nabi, Xiaodong Guo, Mengke Zhang, Mengzhi Wang, Jiaqi Wang, Marie-Laure Fauconnier, Hao Tang, Shujun Bao, Caiming Zhong, Solange Oliveira Rodrigues Valle, Faradiba Serpa Sarquis Serpa, Augusto Sarquis Serpa, María Guadalupe Frías-De León, Eduardo García-Salazar, Gustavo Acosta-Altamirano, Amanda Junior Jorge, Bismark Dabuo, Nunekpeku Xorlali, Ndego Timothy Amoliga, Zyaara Kono Atibodu, Precious Mavis Mavis Newman, Alhassan Mohammed, Raymond Adongsakiya Ali, Abubakari Abudu

First published in London, United Kingdom, 2023 by IntechOpen
IntechOpen is the global imprint of INTECHOPEN LIMITED, registered in England and Wales, registration number: 11086078, 5 Princes Gate Court, London, SW7 2QJ, United Kingdom
Printed in Croatia

British Library Cataloguing-in-Publication Data
A catalogue record for this book is available from the British Library

Additional hard and PDF copies can be obtained from orders@intechopen.com

Aspergillus and Aspergillosis - Advances in Genomics, Drug Development, Diagnosis and Treatment
Edited by Mehdi Razzaghi-Abyaneh, Mahendra Rai and Masoomeh Shams-Ghahfarokhi
p. cm.

This title is part of the Infectious Diseases Book Series, Volume 25
Topic: Fungal Infectious Diseases
Series Editor: Alfonso J. Rodriguez-Morales
Topic Editor: Yuping Ran

Print ISBN 978-1-83768-051-1
Online ISBN 978-1-83768-052-8
eBook (PDF) ISBN 978-1-83768-146-4
ISSN 2631-6188

For EU product safety concerns:
IN TECH d.o.o., Prolaz Marije Krucifikse Kozulić 3, 51000 Rijeka, Croatia,
info@intechopen.com or visit our website at intechopen.com.

IntechOpen Book Series

Infectious Diseases

Volume 25

Aims and Scope of the Series

This series will provide a comprehensive overview of recent research trends in various Infectious Diseases (as per the most recent Baltimore classification). Topics will include general overviews of infections, immunopathology, diagnosis, treatment, epidemiology, etiology, and current clinical recommendations for managing infectious diseases. Ongoing issues, recent advances, and future diagnostic approaches and therapeutic strategies will also be discussed. This book series will focus on various aspects and properties of infectious diseases whose deep understanding is essential for safeguarding the human race from losing resources and economies due to pathogens.

Meet the Series Editor

Dr. Rodriguez-Morales is an expert in tropical and emerging diseases, particularly zoonotic and vector-borne diseases (notably arboviral diseases), and more recently COVID-19 and Monkeypox. He is the president of the Publications and Research Committee of the Pan-American Infectious Diseases Association (API), as well as the president of the Colombian Association of Infectious Diseases (ACIN). He is a member of the Committee on Tropical Medicine, Zoonoses, and Travel Medicine of ACIN. Dr. Rodriguez-Morales is a vice-president of the Latin American Society for Travel Medicine (SLAMVI) and a member of the Council of the International Society for Infectious Diseases (ISID). Since 2014, he has been recognized as a senior researcher at the Ministry of Science of Colombia. He is a professor at the Faculty of Medicine of the Fundacion Universitaria Autonoma de las Americas, in Pereira, Risaralda, Colombia, and a professor, Master in Clinical Epidemiology and Biostatistics, at Universidad Científica del Sur, Lima, Peru. He is also a non-resident adjunct faculty member at the Gilbert and Rose-Marie Chagoury School of Medicine, Lebanese American University, Beirut, Lebanon, and an external professor, Master in Research on Tropical Medicine and International Health, at Universitat de Barcelona, Spain. Additionally, an invited professor, Master in Biomedicine, at Universidad Internacional SEK, Quito, Ecuador, and a visiting professor, Master Program of Epidemiology, at Diponegoro University, Indonesia. In 2021 he was awarded the "Raul Isturiz Award" Medal of the API and, the same year, the "Jose Felix Patiño" Asclepius Staff Medal of the Colombian Medical College due to his scientific contributions to the topic of COVID-19 during the pandemic. He is currently the Editor in Chief of the journal Travel Medicine and Infectious Diseases. His Scopus H index is 55 (Google Scholar H index 77) with a total of 725 publications indexed in Scopus.

Meet the Volume Editors

Professor Mehdi Razzaghi-Abyaneh obtained his Ph.D. in Medical Mycology from Tarbiat Modares University, Tehran, Iran. Dr. Razzaghi-Abyaneh pursued a 12-month sabbatical on the identification of antifungal compounds from bioactive plants at the Graduate School of Agriculture, Tokyo University, in the Laboratory of Applied Biological Chemistry from 2006 to 2007. He is currently a full professor and eminent research scientist at the Pasteur Institute of Iran, where he is working on mycotoxins and mycotoxigenic fungi as well as antifungal nanomaterial, and biologically active antifungals of plant, fungal, and bacterial origin for more than 23 years. He has supervised and advised several Ph.D. and MSc theses. Dr. Razzaghi-Abyaneh has published more than 150 papers in peer-reviewed international journals, 10 books, and several book chapters. His research is focused on the chemical basis of plant–fungal interactions and determining the mode of action of antifungal bioactive compounds of natural origin as small macromolecules at cellular and molecular levels.

Professor Mahendra Rai is a visiting scientist at Nicolaus Copernicus University, Torun, Poland. He has published more than 450 research papers, more than 102 journal articles, and 75 books. He is a member of several scientific societies and has been a national scholar for five years. He received several prestigious awards, including the Father T.A. Mathias Award (1989) from the All India Association for Christian Higher Education, and the Medini Award (1999) from the Department of Environment and Forest, Government of India. He also received a SERC visiting fellowship from the Department of Science and Technology (1996), an INSA visiting fellowship from the Indian National Science Academy (1998), and TWAS-UNESCO Associateship (2002), Italy. Dr. Rai was also awarded a UGC-BSR faculty fellowship by the University Grants Commission, New Delhi (2017–2020) and an NAWA fellowship by the Polish Government (2021–2023). He serves as a referee for twenty international journals and is an editorial board member of ten national and international journals. He has approximately three decades of teaching and research experience. The focus of his research is plant- and nano-based bioactives against human pathogenic microbes.

Professor Masoomeh Shams-Ghahfarokhi is a faculty member at Tarbiat Modares University, Tehran, Iran, where she has been working for around 25 years. After graduating with a degree in Medical Mycology in 2000, Dr. Shams-Ghahfarokhi passed a training course as a Japan Society for the Promotion of Science (JSPS) fellow in Japan during 2005–2007. Her research is focused on the identification and isolation of antifungal compounds from natural resources (including plants and organisms) and their molecular aspects on the activity of pathogenic fungi. She is active in evaluating the inhibitory effects of bio-nanomaterials and anti-scaling membranes for the selective treatment of fungal infections and their mechanism of action on the cellular and molecular levels. She is also investigating the identification of pathogenic fungi and their genetic diversity by using molecular techniques such as sequencing, real-time PCR, multiplex PCR, and other methods. Dr. Shams-Ghahfarokhi has around two decades of teaching and research, and she has advised and supervised numerous MSc and Ph.D. theses.

Contents

Preface

Filamentous fungi of the genus *Aspergillus* reside in soil as the main reservoir and are responsible for life-threatening nosocomial infections with undefined consequences all over the world. There are increasing numbers of such infections due to a rise in immunocompromised patients. At present, around 20 species are known to be involved in the etiology of *Aspergillus*-related diseases under the common name "aspergillosis." Inhalation of airborne conidia is the most important route of infection initiation, and the development of drug resistance against current therapeutic antifungals, especially azoles, is considered the main challenge for clinicians. This book highlights recent advances in *Aspergillus* and aspergillosis from molecular taxonomy and distribution to fungal–host interactions, pathogenesis, immunology, drug development against pathogenic species from bench to clinic, vaccines, novel diagnostic methods, treatment strategies and omics (genomics, proteomics, metabolomics, etc.) approaches to overcome virulence and pathogenicity of etiologic *Aspergillus* species.

The book is divided into five sections comprising 10 chapters. Chapter 1 discusses allergic bronchopulmonary aspergillosis, with emphasis on its clinical outcomes and pathogenesis. Chapter 2 discusses post-viral aspergillosis (PVA) as a clinical form of *Aspergillus* infection after influenza, COVID-19, and the cytomegalovirus infection. Chapter 3 examines the impact of *Aspergillus* infections in patients with selected chronic infections and the treatment of these infections. Chapter 4 shows that *A. fumigatus* requires different strategies by various attributes to infect and cause disease in different hosts and suppress resistance responses by the host. These attributes are influenced by environmental conditions including nutrient composition and response to host defenses. Chapter 5 details the recent progress of nanomaterials-based biosensors to diagnose aspergillosis. Chapter 6 demonstrates that metagenomic next-generation sequencing (mNGS) has the potential to be adopted as the gold standard for the diagnosis of pulmonary aspergillosis, which is a critical step in initiating prompt treatment and improving patients' prognosis. Chapter 7 highlights the importance of antifungal drug resistance against aspergillosis, with a special focus on its outcome in the treatment of clinical cases of the disease. Chapter 8 presents an immune-based nanosensor as a biosensor-Internet of Things (IoT) system as a potential way to achieve real-time detection of aspergillosis. The portable reader serves as an excellent point-of-care tool for routine monitoring of harmful mycotoxins, such as aflatoxins and ochratoxins. Chapter 9 presents an updated review of aflatoxin contamination and its impact and management strategies, providing valuable information for researchers, policymakers, and food safety professionals. Chapter 10 examines the industrial aspect of the genus *Aspergillus*, for example, fermentation of whey protein by *A. niger*, and the importance of released materials.

We would like to thank all authors for their invaluable contributions and hard work. We are also grateful to the staff at IntechOpen, especially Ms. Romina Rovan, who kindly assisted us in coordinating the book and scheduling our activities.

Mehdi Razzaghi-Abyaneh
Department of Mycology,
Pasteur Institute of Iran,
Tehran, Iran

Mahendra Rai
Department of Microbiology,
Nicolaus Copernicus University,
Torun, Poland

Masoomeh Shams-Ghahfarokhi
Faculty of Medical Sciences,
Department of Mycology,
Tarbiat Modares University
Tehran, Iran

Section 1

Advances in *Aspergillus* and Aspergillosis

Chapter 1

Allergic Bronchopulmonary Aspergillosis/Mycosis: An Underdiagnosed Disease

Solange Oliveira Rodrigues Valle, Augusto Sarquis Serpa and Faradiba Sarquis Serpa

Abstract

Allergic bronchopulmonary aspergillosis (ABPA) is an immune-allergic disease of the lung due to a hypersensitivity reaction to antigens of *Aspergillus fumigatus* after colonization into the airways. Predominantly, it affects patients with bronchial asthma and those having cystic fibrosis (CF). Despite being recognized as a distinct entity nearly 70 years ago, this disease remains underdiagnosed. This may be due to the diagnostic methods employed, lack of standardized tests, and diagnostic criteria. The mainstay treatment for ABPA is systemic steroid. Azole antifungal agents represent an alternative for the treatment of exacerbations and are preferential strategy for corticosteroids sparing. Biologic drugs are expected to play an important role in the treatment of ABPA based on their mechanism in inhibition of type 2 inflammation, regulation of eosinophils and IgE levels, and modulation of inflammatory cytokines. Therefore, other studies are necessary for a better understanding of this disease so that an early detection can be done as well as a correct management.

Keywords: allergic bronchopulmonary aspergillosis, *Aspergillus fumigatus*, asthma, cystic fibrosis, diagnostic criteria, bronchiectasis, eosinophilia

1. Introduction

Allergic bronchopulmonary aspergillosis (ABPA) was first identified in 1952 by Dr. K. F.W. Hinson who described eight cases with typical clinical characteristics, including bronchitis/asthma, eosinophilia, bronchiectasis and/or mucus plugs, and isolation of *Aspergillus fumigatus* in lung tissue [1]. In 1968, ABPA was first reported in the United States [2] and 6 years later, in Brazil by our department at Federal University of Rio de Janeiro but remains underdiagnosed to this day, both in Brazil and worldwide [3].

ABPA is a complex pulmonary disorder characterized by an exaggerated hypersensitivity reaction to *Aspergillus* generally *fumigatus* specie, resulting in airway inflammation, mucus plugging, and bronchiectasis [4]. People most at risk of developing ABPA are patients with asthma or cystic fibrosis (CF). Given the propensity of the disease to cause irreversible complications, it is essential to formulate screening protocols for ABPA in these patients. Early detection of the disease is crucial for a better prognosis.

2. Epidemiology

The prevalence of ABPA remains uncertain and is likely to differ across geographical locations. ABPA affects 2.5–15% of patients with asthma and 7–9% of patients with CF, subject to variations within the studied population [5, 6]. In a recent systematic literature review, the pooled prevalence of ABPA in adults with asthma was 11.3%, and in adults, asthmatics sensitized to *A. fumigatus* (*Af*) was 37% [7]. For children, the pooled prevalence of ABPA in subjects with asthma was 9.9% [8]. Studies conducted in our center demonstrated 19% of ABPA prevalence among asthmatics sensitized to *Af* [9] and 12,5% of ABPA prevalence in children diagnosed with CF [10]. Worldwide, more than 4 million people are affected by ABPA [6, 11].

Additionally, genetic predisposition, such as specific HLA-DR alleles, may contribute to the development of ABPA in susceptible individuals [12]. There is no particular age or gender predilection for the occurrence of ABPA, although ABPA is uncommon in patients without asthma or CF [13].

The real prevalence of ABPA is difficult to determine due to the diagnostic methods employed, lack of standardized tests, diagnostic criteria, and the populations studied.

3. Pathogenesis

The pathogenesis of ABPA is complex, involving genetic factors, host–pathogen interactions, hypersensitivity reactions, eosinophilic inflammation, and cytokine dysregulation.

Genetic factors may contribute to the susceptibility to ABPA. Human leukocyte antigen (HLA) genotyping studies have identified specific HLA-DR alleles, such as HLA-DR2 and HLA-DR5, to be associated with an increased risk of developing ABPA in asthmatic and CF patients, respectively [12]. On the other hand, HLA-DQ2 contributes to resistance [14].

Aspergillus is an airborne ubiquitous saprophytic fungus that is found in soil and grows on decaying vegetation [15]. Spores are small (2–3 mm), facilitating their deposition throughout the airways. Inhalation of *Aspergillus* spores can lead to colonization and germination in the airways of susceptible individuals, such as those with asthma or CF [16]. *Aspergillus* spores persist in lower airways and develop the ability to germinate into mycelial filaments. This results in the secretion of metabolites that drive the activation of mucosal innate immune response and exposition of *Aspergillus* to the immune system. The impaired mucociliary clearance and local immune response in these individuals facilitate the persistence of *Aspergillus* hyphae in the airways, triggering an exaggerated immune response [17].

ABPA is characterized by a combination of type I (immediate) and type III (immune complex-mediated) hypersensitivity reactions [16]. In type I hypersensitivity, *Aspergillus* antigens bind to immunoglobulin E (IgE) on the surface of mast cells and basophils, leading to the release of inflammatory mediators such as histamine, leukotrienes, and prostaglandins, causing bronchoconstriction, increased vascular permeability, and mucus production [18]. In type III hypersensitivity, immune complexes composed of *Aspergillus* antigens and immunoglobulins (IgG) deposit in the lung tissue, activating complement and attracting inflammatory cells, leading to tissue damage and eosinophilic inflammation [19].

Eosinophils play a central role in the pathogenesis of ABPA, contributing to airway inflammation, mucus production, and bronchial hyperresponsiveness [16].

There is a strong type-2 (T2) inflammation in ABPA combining a massive infiltration of airways by eosinophils and a high level of polyclonal IgE. Cytokines, such as interleukin IL-4, IL-5, and IL-13, produced by Th2 cells, stimulate eosinophil recruitment, activation, and survival [20]. Additionally, IL-4 and IL-13 promote the production of IgE by B cells and induce goblet cell hyperplasia, increasing mucus secretion. Airway inflammation leads to production of dense eosinophil mucus containing Charcot–Leyden crystals that obstructs airways. Mutation in CF transmembrane conductance regulator (CFTR) may play a role in ABPA even in the absence of CF, although this is still not clear [21].

4. Clinical manifestation

The clinical presentation of ABPA is heterogeneous, ranging from mild to severe respiratory symptoms. The primary clinical manifestations of ABPA are respiratory symptoms, which are often similar to those of asthma or CF, making the differentiation between these conditions difficult [22]. Common respiratory symptoms of ABPA include paroxysmal episodes of coughing, wheezing, dyspnea, and chest tightness [23]. Patients with ABPA may also experience recurrent episodes of pulmonary exacerbations, characterized by worsening respiratory symptoms and increased sputum production [4].

A characteristic feature of ABPA is the expectoration of brownish-black mucus plugs, which contain fungal hyphae and eosinophilic material [23]. These mucus plugs can lead to airway obstruction and impaired mucociliary clearance, contributing to the development of bronchiectasis and recurrent infections [22].

Systemic symptoms, such as fever, malaise and weight loss, may occur in ABPA, particularly during exacerbations [23]. These symptoms are thought to be related to the release of inflammatory mediators and the immune response to *Aspergillus* antigens [4].

Evolution is marked by the occurrence of ABPA exacerbations characterized by the onset, or an increase in, the clinical manifestations of ABPA often associated with radiological abnormalities and elevation of eosinophils and total IgE.

5. Diagnostic criteria

ABPA occurs mainly in patients diagnosed with asthma or CF. The diagnosis is based on a combination of clinical manifestations, radiological, and immunological features [24]. A thorough examination of the existing literature highlights the importance of early recognition and appropriate management of ABPA to prevent disease progression and associated complications. Over the years, the diagnostic criteria and approaches for ABPA have evolved significantly (**Table 1**).

The first diagnostic criteria described for ABPA were primarily based on clinical features, blood eosinophilia, and radiological findings [1]. In the 1970s, after the description of the IgE antibody isotype, serology came to occupy a central role in diagnosing ABPA. Rosenberg and Patterson proposed two groups of criteria to improve diagnostics. These criteria (seven primary, three secondary) have since been the most used in the diagnosis of ABPA [25].

ABPA Criteria	
Rosenberg–Patterson (1977) [25]	Primary criteria 1. Asthma 2. Peripheral blood eosinophilia 3. Immediate skin reactivity to *Aspergillus* antigen 4. Precipitating antibodies to *Aspergillus* antigen 5. 5 5. Elevated serum IgE concentration 6. History of pulmonary infiltrates (transient or fixed) 7. Central bronchiectasis Secondary criteria • *Aspergillus fumigatus* in sputum • History of expectoration of brown plugs • Late skin reactivity to *Aspergillus* antigen (Arthus reaction) ABPA "likely": primary criteria 1 to 6 are present. ABPA "certain": all primary criteria are present.
Modified ISHAM criteria (2013) [26]	Predisposing condition • Asthma or cystic fibrosis Obrigatory criteria • Total IgE level > 500 kIU/l • Serum IgE against *Aspergillus fumigatus* > 0,35 kUA/l or positive skin test Other criteria • Blood eosinophilia >500 cels/μL * • Precipitins or increased IgG antibody to *Aspergillus* • CT scan showing bronchiectasis • Mucus impactation on CT scan Obrigatory criteria: both must be presented Other criteria: at least two must be presented, *without systemic corticosteroid
Modified ISHAM criteria (2021) [27]	Predisposing condition • Asthma Obrigatory criteria • Total IgE level > 500 kIU/l • Serum IgE level against *Aspergillus fumigatus* > 0,35 kUA/l or positive skin test Other criteria • Blood eosinophilia >500 cels/μL • Precipitins or increased IgG antibody to *Aspergillus* > 27 mg_A/l • CT scan showing bronchiectasis • Mucus impactation on CT scan Obrigatory criteria: both must be present Other criteria: at least 2 must be present
Asano Criteria (2021) [28]	For Allergic Bronchopulmonary Mycosis 1. Asthma 2. Blood eosinophilia >500/μL 3. Total IgE level > 417 kIU/l 4. Positive immediate skin test or specific IgE >0,35 kUA/l for filamentous fungi 5. Growth of filamentous fungi in culture of sputum or bronchial lavage fluid 6. Presence of fungal hyphae in mucus plugs 7. Central bronchiectasis on CT scan 8. Presence of plugs on CT scan/bronchoscopy, or history of mucus plug expectoration 9. High attenuation mucus on CT scan Diagnosis ABPM: At least six criteria must be present

Table 1.
Diagnostic criteria for ABPA.

However, these criteria lacked sensitivity and specificity, leading to underdiagnosis or misdiagnosis. In 2013, the ABPA Working Group of the International Society for Human and Animal Mycology (ABPA-ISHAM) proposed new diagnostic criteria to provide standardized guidelines for ABPA diagnosis [26]. New evidence then came to light about the "sensibility and specificity" of the ISHAM criteria. This evidence emerged through a study that used a latent class analysis to explore the performance of various existing and novel diagnostic criteria [27]. It has been demonstrated that IgE-specific tests are more responsive than skin tests to identify sensitization to *Aspergillus* and that the sensitivity and specificity of these criteria increased using a threshold for total IgE at 500 kIU/l [27]. In addition, a *cut-off* for specific IgG to *Aspergillus* was proposed. This evidence led the ISHAM to modify the diagnostic criteria. These modifications offered improved diagnostic performance and were published in 2021 [27]. The criteria proposed by ISHAM were validated but revealed that the sensitivity was poor for cases with non-*Aspergillus* Allergic Bronchopulmonary Mycosis (ABPM). Asano et al. proposed and validated new diagnostic criteria that showed improved sensitivity and specificity compared to Rosenberg-Paterson and ISHAM's previous criteria, even in atypical cases without asthma or non-*Aspergillus* ABPM (**Table 1**) [27].

The diagnostic of ABPA often follows an algorithm that integrates multiple pieces of clinical, immunological, and radiological information. The starting point is a clinical assessment, looking for key symptoms and risk factors, especially asthma or CF and then elevated specific IgE to *Af*. Total serum IgE levels are also evaluated, which typically exceed 500 IU/mL, while peripheral eosinophils >500 cels/μL and specific IgG-*Af* > 27 mg_A/l provide further evidence of a patient's immunological response to the fungus [27]. Chest high-resolution computed tomography (HRCT) is another essential component of the algorithm because it provides a classification of ABPA based on the features observed (**Figure 1**).

While recent updates to diagnostic criteria have led to some advances, ABPA remains a challenging condition to diagnose in clinical practice. It is important to note that the diagnostic algorithm may vary slightly based on specific guidelines and updates in diagnostic technology, for example, the determination of specific IgE levels indicative of ABPA is subject to debate. Patterson et al. in 1983 had already described that the levels of specific IgE and IgG for *Af* were twice as high in patients with ABPA as compared to patients with Af-sensitive asthma [29]. Employing a cutoff of 0.35 kUA/L for *Af*-IgE could potentially result in overdiagnosis of ABPA [30]. Furthermore, the cutoff to IgG-Af proposed in ISHMA criteria was determined in Asian population using the ImmunoCAP® method and is much lower than values reported by other European studies in ABPA patients [31].

Recently, the application of molecular allergology has been proposed. These tests evaluate IgE-directed against alergens components of *Af* (r*Asp*) to improve diagnosis taking into account that *Af*-specific IgE cannot distinguish sensitization from allergy to *Af* [32]. It is well established that specific proteins of *Af* could play a significant role in ABPA, and these can be detected through this approach. The presence of either r*Asp* f1 or r*Asp* f3 demonstrated high sensitivity and r*Asp* f4 or r*Asp* f6 showed high specificity in diagnosing ABPA in patients with asthma and CF [33]. Based on the proposal of a diagnostic algorithm that includes r*Asp* to improve the accuracy of diagnosis [32], the EAACI ABPA Task Force proposed changes to the algorithm and reinforced the need to adequate recommendations for countries with limited resources (**Figure 2**)[30].

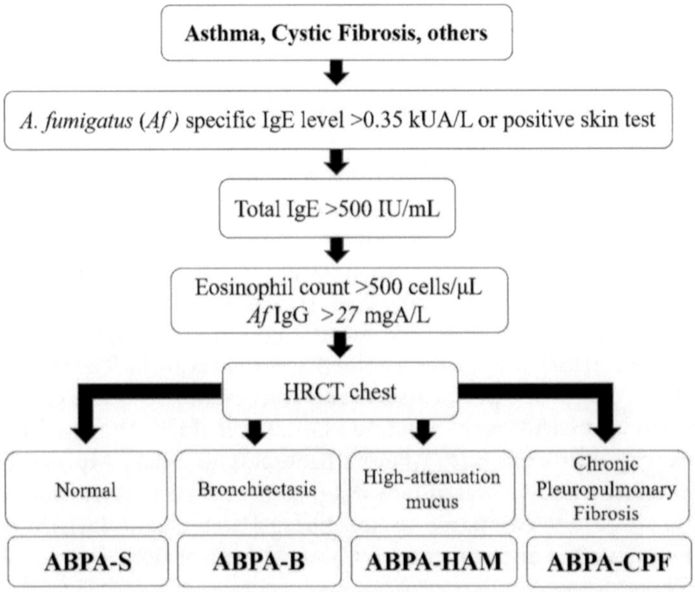

Adaptado de Agarwal et al. Expert Review of Respiratory Medicine. 2016;10(12), 1317–34.

Figure 1.
Algorithm to diagnosis and radiographic classification of ABPA. Af: Aspergillus *fumigatus; HCTR: Highresolution computed tomography; ABPA-S: Serologic ABPA; ABPA-B: ABPA with bronchiectasis; ABPA-HAM: ABPA with high-attenuation mucus; ABPA-CPF: ABPA with chronic pleuropulmonary fibrosis.*

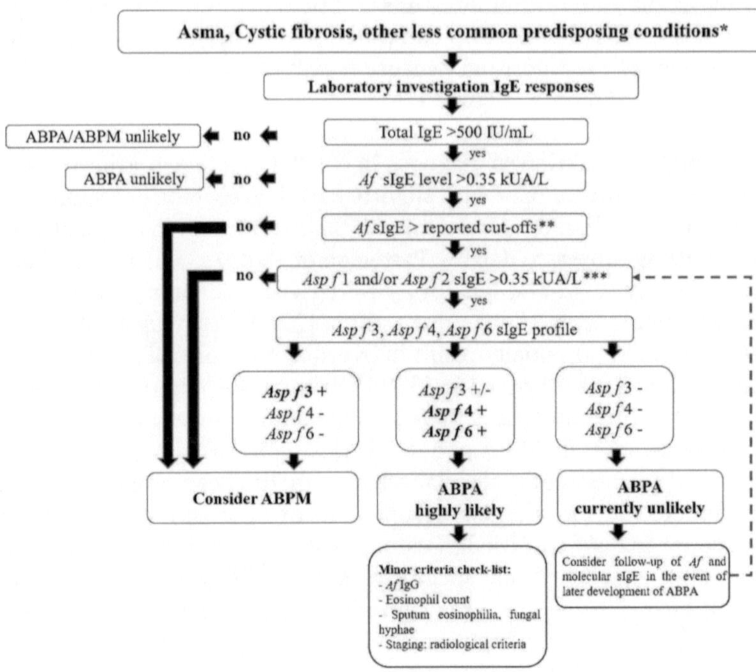

Adaptado de Carminati M. et al. Allergy. 2022;77:3476–3477.

Figure 2.
Laboratorial investigation of recombinant Aspergillus *antigen-based algorithm for diagnosing ABPA. *chronic obstructive pulmonary disease, altered bronchopulmonary structure; **consider the establishment of locally validated cut-offs; ***consider the ratio of sIgE to molecular allergens vs. Af sIgE.*

Chest radiographic imaging plays a vital role in the diagnosis and management of ABPA and may exhibit features, ranging from normal to manifestations of pulmonary fibrosis (**Table 2**) [34]. This spectrum of presentations reflects the potential progression of the disease and the variability of the individual's response to *Aspergillus*. Several studies have investigated these radiographic features [34–36]. HRCT was demonstrated to be the gold standard for detecting bronchiectasis distribution and in identifying subtle radiographic changes, such as tree-in-bud opacities, which may indicate small airway involvement in ABPA [36, 37].

In the acute stage of ABPA, transient pulmonary infiltrates may be observed, typically manifesting as patchy opacities or consolidation [26, 34–36]. These infiltrates may resolve spontaneously or with treatment.

As ABPA progresses, more prominent, persistent changes can be detected in HRCT. These include central bronchiectasis, predominantly in the upper lobes [35]. High-attenuation mucus (HAM) within bronchi, often associated with bronchial dilations resembling a "finger-in-glove" sign, is another critical feature [26]. HAM is said to be present when the density of mucus is visibly greater than that of the paraspinal muscle, and that it may be related to the presence of calcium salts and metal ions (manganese or iron) [38] or desiccated mucus [39]. Chronic stage indicators also encompass bronchial wall thickening and parenchymal scarring, reflective of long-standing inflammation and damage [34–36]. Centrilobular nodules with a "tree-in-bud" pattern, indicative of small airway inflammation, often associated with chronic inflammation, are commonly identified in ABPA [37]. The end stage of the disease is fibrotic and can be identified by the presence of fibrosis and architectural distortion, predominantly involving upper lobes. Pulmonary fibrosis can manifest as traction bronchiectasis, honeycombing, and volume loss on HRCT [34, 37].

Chest radiography findings	
Transitory	• Peripheral infiltrate • Air-fluid levels • Consolidations • Toothpaste image • Train line image • Lobe or segmental collapse
Permanent	• Bronchiectasis • Parallel line image • Ring shadow • Pulmonary fibrosis
High-resolution chest tomography findings	
Bronchi	• Bronchiectasis • Mucus plug with high attenuation • Dilated bronchi • Occluded bronchi • Bronchial thickening
Parenchyma	• Cavitation • Bullous emphysema • Pleural involvement • Pneumothorax • Pleural fibrosis

Table 2.
Radiographics features described in ABPA.

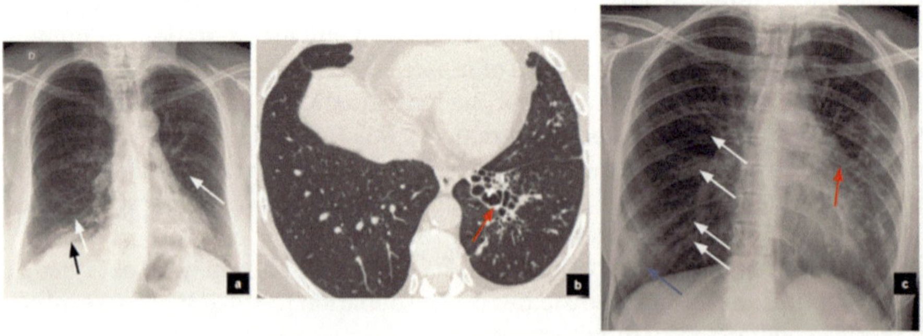

Figure 3.
(a) A 64-year-old female. Radiograph with consolidation in the right lower lobe (black arrow), in addition to some bilateral bronchiectasis with parietal thickening (white arrows). (b) Chest CT of the same patient demonstrates multiple central bronchiectasis in the left lower lobe with thickening and interspersed hyperdense content, a finding highly specific for ABPA (allergic bronchopulmonary aspergillosis). (c) a 16-year-old female. Radiograph shows consolidations in the peripheral region of the left lung (red arrow) and in the right lower lobe (blue arrow), in addition to bronchiectasis with thickened walls, the tram-track sign (white arrows).

The nuanced insights that HRCT offers into the extent and characteristics of ABPA make it a key tool in the image-based classification of this disease [26, 35].

Using HRCT, ABPA can be classified into four types or phenotypes. Patients showing no abnormalities on chest scan are classified as having serologic ABPA (ABPA-S) [26, 40]. In contrast, those with evidence of central bronchiectasis are labeled as having ABPA Central Bronchiectasis (ABPA-CB) [26, 40]. The presence of HAM leads to the categorization of the disease as ABPA-High Attenuation Mucus (ABPA-HAM) [36, 40]. Lastly, if at least two radiological features suggestive of fibrosis (fibrocavitary lesions, pulmonary fibrosis, and pleural thickening) are observed in the absence of mucoid impaction (or HAM), the disease is classified as ABPA-Chronic Pleuropulmonary Fibrosis (ABPA-CPF) (**Figures 1**, **3**, and **4**) [36, 40].

Timely recognition and appropriate monitoring of radiographic changes are essential for optimizing patient care and outcomes in ABPA.

6. Staging of ABPA

Staging of ABPA is an important aspect of its diagnosis, management, and understanding of the disease's progression. These stages were initially proposed by

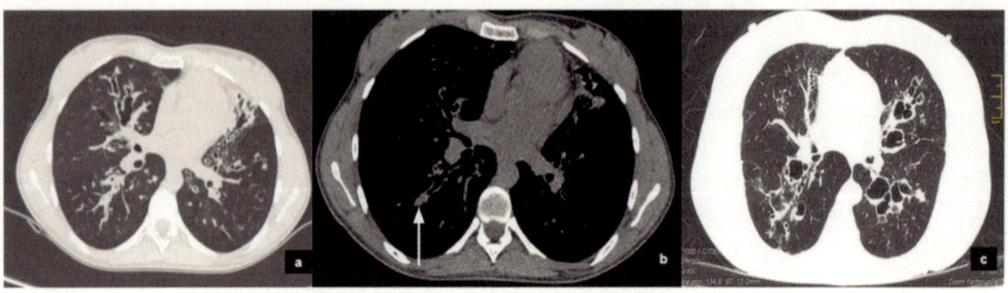

Figure 4.
CT imaging of patients diagnosed with ABPA: (a) multiple bilateral bronchiectasis with parietal thickening and mucoid impactation; (b) mucoid impactation in mediastinal window of the same patient (white arrow); (c) central bronchiectasis in a corticosteroid-dependent patient who has experienced multiple exacerbations.

Stage	Clinical, radiographic, and laboratorial features
0 - Asymptomatic	• No previous diagnosis of ABPA • Controlled asthma • Fulfilling the diagnostic criteria of ABPA
1 - Acute	• No previous diagnosis of ABPA • Uncontrolled asthma/symptoms consistent with ABPA • Satisfying the diagnostic criteria of ABPA
1a - With mucoid impaction	• Mucoid impaction observed on thoracic imaging
1b - Without mucoid impaction	• Absence of mucoid impaction on thoracic imaging
2 - Remission	• Clinical and/or radiological improvement and • Decline in serum total IgE by ≥25% of baseline at 8 weeks
3 - Exacerbation	• Clinical and/or radiological worsening and • Increase in serum total IgE by at least 50% from the new baseline established during response/remission
4 - Remission	• Sustained clinical and radiological improvement and • Serum total IgE levels persisting at or below baseline (or increase by <50%) for ≥6 months off treatment
5a - Treatment-dependent ABPA	• Two or more exacerbations within 6 months of stopping therapy OR • Clinical and/or radiological worsening, along with increase in serum total IgE levels, on tapering oral steroids/azoles
5b – Steroid dependent asthma	• Systemic corticosteroids required for control of asthma while the ABPA activity is controlled (as indicated by serum total IgE and thoracic imaging)
6 - Advanced ABPA	• Extensive bronchiectasis due to ABPA on chest imaging AND • Complications (cor pulmonale and/or chronic Type II respiratory failure

Adapted from Agarwal et al. [24]. Ref. [40].

Table 3.
Staging of ABPA.

Rosenberg et al. and have been widely used [25]. The staging has been revisited and revised by Agarwal et al., who proposed a more detailed, six-stage system, which also subdivides some stages based on radiological findings (**Table 3**) [26, 40].

7. Differential diagnosis

The differential diagnosis of ABPA includes several pulmonary disorders that share clinical and radiological features with ABPA, such as asthma with *Af* sensitization, idiopathic chronic eosinophilic pneumonia, tuberculosis, nontuberculous mycobacterial infections, eosinophilic granulomatosis with polyangiitis (GEPA) [4]. Distinguishing ABPA from these conditions requires a thorough clinical assessment, including the identification of predisposing factors, and appropriate tests to confirm or exclude these conditions.

8. Treatment

ABPA is recognized as a treatable trait in patients with bronchiectasis and the expected benefits of treatment are the prevention of lung damage, improved outcome and quality of life [41]. To achieve these goals, the management of ABPA varies depending on the stage of the disease and involves a combination use of systemic corticosteroid, antifungal therapy, immunobiological agents, and airway clearance techniques tailored to individual patient needs (**Table 4**).

Drugs		Indication
Oral corticosteroids*	• Prednisolone (or equivalent), 0.5 mg/kg over a period of 14 days, followed by 0.5 mg/kg/day on alternate days for 8 weeks, then reduce 5 mg every 2 week. Total duration of 3–5 months OR • Prednisolone (or equivalent), 0.75 mg/kg/day for 6 weeks, then reduce 5 mg every 6 weeks. Total duration: 6–12 months	• Acute stage • Exacerbation stage
Intravenous corticosteroid	• Methylprednisolone, 15 mg/day (note exceeding 1 g) intravenous for 3 consecutive days	• To minimize the side effects of daily corticosteroid therapy • Refractory ABPA exacerbations
Antifungal	Oral antifungal therapy:† • Itraconazol 200 mg twice a day for 16 weeks OR • Voriconazole 200 mg twice a day, for at least 24 weeks Nebulized antifungal therapy: • Amphotericin B§	• Acute stage • Exacerbation stage
Immunobiological#	• Omalizumab 150 mg: approved for ages ≥6 years, SC injection. Dose and frequency depend on the patient's weight and the serum IgE level at the start of treatment. Highest permitted serum total IgE is 1500 IU/mL OR • Mepolizumab 40 mg or 100 mg: approved for ages ≥6 years. For children 6–11 years, 40 mg SC injection every 4 weeks. For ≥12 years, 100 mg SC injection every 4 weeks OR • Benralizumab 30 mg: approved for ages ≥12 years, SC injection every 4 weeks for three doses then every 8 weeks OR • Dupilumab 200 mg or 300 mg: approved for ages ≥6 years, SC injections every 2 weeks. For children 6–11 years, dose and frequency depend on weight. For ages ≥12 years 300 mg for asthma corticosteroid-dependent severe asthma or concomitant moderate/severe atopic dermatitis OR • Tezepelumab 210 mg: approved for ages ≥12 years, SC injection every 4 weeks	• Refractory ABPA • Uncontrolled asthma • Adverse effects or • Contraindications to corticosteroids and azoles

**First-line treatment. For treating the first exacerbation, corticosteroids could be used alone and combined with azoles for subsequent exacerbation.*
†Itraconazol could be used alone in those with contraindications to corticosteroids. The associated use with prednisolone may reduce the chance of exacerbations.
#Randomized Clinical Trials evaluating immunobiologicals are necessary to clarify the role of these agents in treatment of ABPA.
§Nebulized amphotericin B may be considered when prolonged use of systemic corticosteroids and/or azoles are necessary.

Table 4.
Drugs used in the management of ABPA.

8.1 Corticosteroids

Oral corticosteroids are the first-line treatment for ABPA [42]. However, the ideal dosage and duration of treatment remain undefined, and there are several treatment regimens, each featuring varying dosages and durations of use [40]. The most common regimen begins with a daily administration of 0.5 mg/kg of prednisolone over a period of 14 days, followed by 0.5 mg/kg/day on alternate days for 8 weeks, then taper by 5 mg every 2 weeks to complete a total steroid duration of 3–5 months [23].

Although disease remission can be achieved in most cases treated with medium- to high-dose systemic corticosteroids, relapse occurs in a substantial proportion of patients (13.5–45%) and can become corticosteroid-dependent [42, 43].

Intravenous pulse dose of corticosteroids can be used as a substitute for oral administration. Methylprednisolone, 15 mg/day (note exceeding 1 g), has been used in children to minimize the side effects of daily corticosteroid therapy [44] and in cases of refractory ABPA exacerbations [45].

8.2 Antifungal therapy

Antifungal agents decrease the fungal burden in the airways, antigenic stimulus, inflammatory response, and can contribute to reducing exposure to systemic corticosteroids. The use of azole agents alone or in combination with corticosteroid is an option in the treatment of ABPA [46]. Azoles, such as itraconazole, is generally prescribed at a dose of 200 mg twice daily for 16 weeks [40]. The combined use of itraconazole and prednisolone resulted in a greater reduction in the one-year exacerbation rate when compared to the use of these drugs individually [46].

However, drug interactions, hepatotoxicity, and variable bioavailability may limit the use [4]. Other azoles, such as voriconazole and posaconazole, have also been effective in treating ABPA, especially in cases of itraconazole intolerance or resistance [47].

The efficacy of nebulized amphotericin B in the management of ABPA exacerbations seems to be limited. However, they may be considered when other alternative options have been exhausted [24].

8.3 Immunobiological drugs

Posttreatment recurrences of ABPA are commonly seen, whether using oral corticosteroids, antifungal therapy or a combination of both, and prolonged treatment can result in adverse effects. Therefore, the necessity for new, safe, and effective treatment strategies is clear. Given the pathogenesis of ABPA, biologics designed to target type 2 inflammation, initially developed for severe asthma management, are expected to potentially serve as effective treatment alternatives for ABPA [48]. Although limited by the scarcity of randomized controlled trials, recent case reports and case series have demonstrated the benefits of target type 2 inflammation in the treatment of ABPA [48–51].

The anti-IgE monoclonal antibody, omalizumab, has shown potential in reducing corticosteroid use, improving lung function, and preventing relapses [50]. Omalizumabe is administered subcutaneously every 2–4 weeks, with the dosage determined based on the patient's weight and baseline serum total IgE levels [50]. However, the doses used might be suboptimal due to the high levels of IgE observed in ABPA.

The two groups of biologics target IL-5/eosinophil pathway: monoclonal antibodies that target IL-5, like mepolizumab and reslizumab, and those against the IL-5 receptor-alpha chain (IL-5Rα), such as benralizumab, have demonstrated efficacy in managing resistant eosinophilic pulmonary disorders, including ABPA. These anti-IL-5/IL-5Rα mAbs have been successful in reducing exacerbation frequency, dosage of oral corticosteroids, and enhancing pulmonary function in patients with asthma-complicated ABPA, even those unresponsive to omalizumab [48].

Dupilumab, anti-IL-4Rα monoclonal antibody, have been shown therapeutic effects on the symptoms and pulmonary function [49, 52]. Some patients with ABPA refractory to treatment with omalizumab or mepolizumab responded to dupilumab treatment [49, 53].

Tezepelumab a human IgG2 monoclonal antibody that binds specifically to thymic stromal lymphopoietin (TSLP) was demonstrated to improve the control of severe asthma by normalizing broad inflammatory pathways [54]. A recently published case report on a patient using mepolizumab, showed benefits in control of symptoms and reduction of the mucus plugs and pulmonary opacities [51].

Airway clearance techniques, such as chest physiotherapy and positive expiratory pressure devices, may be beneficial in patients with ABPA, particularly those with coexisting CF [4]. These techniques can help remove mucus plugs, improve lung function, and reduce the risk of recurrent infections.

The treatment of ABPA in CF is not very different from that of ABPA in asthma. As patients with CF often have coexisting malabsorption, treatment is more complex as oral medications, especially itraconazole capsules may be poorly absorbed [4].

8.4 Monitoring of treatment

The response to treatment should be monitored with clinical parameters, chest radiograph, and measurements of the serum total IgE concentration every 8 weeks. There should be a resolution of radiographic opacities and a 25% minimum reduction in serum total IgE levels and it is necessary to establish the "new" baseline level [4]. Clinical and/or radiological worsening along with 50% increase in IgE levels suggests an exacerbation [4].

9. Conclusion

ABPA is an immune-allergic disease of airways occurring in genetically predisposed patients as asthma and CF. The exact prevalence is not yet well known, and range is quite extensive as there is no single clinical, radiological, or serological parameter to make the diagnosis, leading to the use of various diagnostic criteria. Due to the absence of a consensus, ABPA may be easily underdiagnosed. Therapeutic management is based on few controlled studies conducted in asthma and extrapolated to ABPA. The first line of treatment of exacerbations remains on use of oral corticosteroids. Azole antifungal agents represent an alternative for the treatment of exacerbations and are preferential strategy for corticosteroids sparing. Asthma biologics may be a potential pharmacological management in the future. Therefore, more studies are needed regarding the diagnostic and therapeutic criteria for a better management of these patients.

Conflict of interest

The authors declare that they have no conflict of interest regarding this work.

Author details

Solange Oliveira Rodrigues Valle[1]*, Augusto Sarquis Serpa[2]
and Faradiba Sarquis Serpa[3]

1 Department of Internal Medicine, Immunology Service, Clementino Fraga Filho University Hospital, Federal University of Rio de Janeiro, Rio de Janeiro, Brazil

2 Federal University of São Paulo, Diagnostic Imaging Department, São Paulo, Brazil

3 Santa Casa de Misericórdia de Vitória School of Sciences, Asthma Reference Center, Espírito Santo, Brazil

*Address all correspondence to: solangervalle@gmail.com

References

[1] Hinson KFW, Moon AJ, Plummer NS. Broncho-pulmonary aspergillosis: A review and a report of eight new cases. Thorax. 1952;7(4):317-333

[2] Patterson R, Golbert TM. Hypersensitivity disease of the lung. University of Michigan Medical Center Journal. 1968;**34**(1):8-11

[3] França AT. Asma Brônquica: importância do Aspergillus fumigatus. Tese de Livre Docência. Rio de Janeiro: Universidade Federal do Rio de Janeiro; 1974

[4] Agarwal R, Sehgal IS, Dhooria S, Muthu V. Allergic bronchopulmonary aspergillosis. Chest. 2020;**157**(2): 403-416

[5] Agarwal R, Aggarwal AN, Gupta D, Jindal SK. *Aspergillus* hypersensitivity and allergic bronchopulmonary aspergillosis in patients with bronchial asthma: Systematic review and meta-analysis. The International Journal of Tuberculosis and Lung Disease. 2009;**13**: 936-944

[6] Denning DW, Pleuvry A, Cole DC. Global burden of allergic bronchopulmonary aspergillosis with asthma and its complication chronic pulmonar aspergillosis in adults. Medical Mycology. 2013;**51**:361-370

[7] Agarwal R, Muthu V, Sehgal IS, Dhooria S, Prasad KT, Soundappan K, et al. Prevalence of aspergillus sensitization and allergic bronchopulmonary aspergillosis in adults with bronchial asthma: A systematic review of global data. The Journal of Allergy and Clinical Immunology: In Practice. 2023;11(6): 1734-1751

[8] Agarwal R, Muthu V, Sehgal IS, Dhooria S, Prasad KT, Soundappan K, et al. Aspergillus sensitization and allergic bronchopulmonary aspergillosis in asthmatic children: A systematic review and meta-analysis. Diagnostics (Basel). 2023;**13**(5):922

[9] Serpa FS. Aspergilose Broncopulmonar Alérgica: critérios diagnósticos em pacientes sensíveis ao Aspergillus fumigatus. UFRJ: Tese de Mestrado; 1997

[10] Valle SOR. Identificação de Aspergilose broncopulmonar alérgica nos pacientes com FC e reatividade cutânea aos antígenos do Af. UFRJ: Tese de Mestrado; 1998

[11] Gago S, Denning DW, Bowyer P. Pathophysiological aspects of aspergillus colonization in disease. Medical Mycology. 2019;**57**(Supplement_2):S219-S227

[12] Chauhan B, Santiago L, Kirschmann DA, Hauptfeld V, Knutsen AP, Hutcheson PS, et al. The association of HLA-DR alleles and T cell activation with allergic bronchopulmonary aspergillosis. Journal of Immunology. 1997;**159**:4072-4076

[13] Muthu V, Sehgal IS, Prasad KT, Dhooria S, Aggarwal AN, Garg M, et al. Allergic bronchopulmonary aspergillosis (ABPA) sans asthma: A distinct subset of ABPA with a lesser risk of exacerbation. Medical Mycology. 2020;**58**:260-263

[14] Chauhan B, Santiago L, Hutcheson PS, Schwartz HJ, Spitznagel E, Castro M, et al. Evidence for the involvement of two different MHC class II regions in susceptibility or protection in allergic bronchopulmonary

aspergillosis. The Journal of Allergy and Clinical Immunology. 2000;**106**(4): 723-729

[15] Van de Veerdonk FL, Gresnigt MS, Romani L, Netea MG, Latgé JP. Aspergillus fumigatus morphology and dynamic host interactions. Nature Reviews Microbiology. 2017;**15**(11): 661-674

[16] Knutsen AP, Bush RK, Demain JG, Denning DW, Dixit A, Fairs A, et al. Fungi and allergic lower respiratory tract diseases. The Journal of Allergy and Clinical Immunology. 2012;**129**(2): 280-291 quiz 292-3

[17] Chaudhary N, Datta K, Askin FB, Staab JF, Marr KA. Cystic fibrosis transmembrane conductance regulator regulates epithelial cell response to aspergillus and resultant pulmonary inflammation. American Journal of Respiratory and Critical Care Medicine. 2012;**185**:301-310

[18] Knutsen AP, Slavin RG. Allergic bronchopulmonary aspergillosis in asthma and cystic fibrosis. Clinical & Developmental Immunology. 2011;**2011**: 843763

[19] Hohl TM, Feldmesser M. Aspergillus fumigatus: Principles of pathogenesis and host defense. Eukaryotic Cell. 2007; **6**(11):1953-1963

[20] Roboubi A, Audousset C, Fréalle É, Brun AL, Laurent F, Vitte J, et al. Allergic bronchopulmonary aspergillosis: A multidisciplinary review. Journal of Medical Mycology. 2023;**33**(3): 101392

[21] Agarwal R, Khan A, Aggarwal AN, Gupta D. Link between CFTR mutations and ABPA: A systematic review and meta-analysis. Mycoses. 2012;**55**(4): 357-365

[22] Maturu VN, Agarwal R. Prevalence of aspergillus sensitization and allergic bronchopulmonary aspergillosis in cystic fibrosis: Systematic review and meta-analysis. Clinical and Experimental Allergy. 2015;**45**(12):1765-1778

[23] Greenberger PA. Allergic bronchopulmonary aspergillosis. The Journal of Allergy and Clinical Immunology. 2002;**110**(5):685-692

[24] Agarwal R, Sehgal IS, Dhooria S, Aggarwal AN. Developments in the diagnosis and treatment of allergic bronchopulmonary aspergillosis. Expert Review of Respiratory Medicine. 2016; **10**(12):1317-1334

[25] Rosenberg M, Patterson R, Mintzer R, Cooper BJ, Roberts M, Harris KE. Clinical and immunologic criteria for the diagnosis of allergic bronchopulmonary aspergillosis. Annals of Internal Medicine. 1977;**86**(4): 405-414

[26] Agarwal R, Chakrabarti A, Shah A, Gupta D, Meis JF, Guleria R, et al. Allergic bronchopulmonary aspergillosis: Review of literature and proposal of new diagnostic and classification criteria. Clinical and Experimental Allergy. 2013; **43**(8):850-873

[27] Saxena P, Choudhary H, Muthu V, Sehgal IS, Dhooria S, Prasad KT, et al. Which are the optimal criteria for the Diagnosis of allergic bronchopulmonary aspergillosis? A latent class analysis. The Journal of Allergy and Clinical Immunology. In Practice. 2021;**9**(1): 328-335.e1

[28] Asano K, Hebisawa A, Ishiguro T, Takayanagi N, Nakamura Y, Suzuki J, et al. New clinical diagnostic criteria for allergic bronchopulmonary aspergillosis/ mycosis and its validation. The Journal of

Allergy and Clinical Immunology. 2021; **147**(4):1261-1268.e5

[29] Patterson R, Greenberg PA, Ricketti AJ, Roberts MA. Radioimmunoassay index for allergic bronchopulmonary aspergillosis. Annals of Internal Medicine. 1983;**99**:18-22

[30] Caminati M, Feleszko W, Michel M, Annesi-Maesano I, Vitte J, Task Force Diagnosis EAACI, et al. Aspergillus fumigatus and personalized medicine: Toward a clinically reliable algorithm. Allergy. 2022;**77**(11):3476-3477

[31] Raulf M, Marcus J, Ingrid S, Hoffmeyer F, Nowak D, Ochmann U, et al. Update of reference values for IgG antibodies against typical antigens of hypersensitivity pneumonia. Allergo Journal International. 2019;**28**:192-206

[32] Lukaszewicz R, Mahay G, Boyer O, Martinet J. Medical algorithm: Aspergillus fumigatus components in the diagnosis of allergic bronchopulmonary aspergillosis. Allergy. 2022;**77**(1):327-330

[33] Muthu V, Sehgal IS, Dhooria S, Aggarwal AN, Agarwal R. Utility of recombinant aspergillus fumigatus antigens in the diagnosis of allergic bronchopulmonary aspergillosis: A systematic review and diagnostic test accuracy meta-analysis. Clinical and Experimental Allergy. 2018;**48**: 1107-1136

[34] Agarwal R, Khan A, Garg M, Aggarwal AN, Gupta D. Chest radiographic and computed tomographic manifestations in allergic bronchopulmonary aspergillosis. World Journal of Radiology. 2012;**4**(4):141-150

[35] Agarwal R, Gupta D, Aggarwal AN, Saxena AK, Chakrabarti A, Jindal SK. Clinical significance of hyperattenuating mucoid impaction in allergic bronchopulmonary aspergillosis: An analysis of 155 patients. Chest. 2007 Oct; **132**(4):1183-1190

[36] Agarwal R, Khan A, Gupta D, Aggarwal AN, Saxena AK, Chakrabarti A. An alternate method of classifying allergic bronchopulmonary aspergillosis based on high-attenuation mucus. PLoS One. 2010;**5**(12):e15346

[37] Agarwal R, Khan A, Gupta D, et al. An official ATS/ERS/JRS/ALAT statement: HRCT pattern of lung disease in chronic hypersensitivity pneumonitis. American Journal of Respiratory and Critical Care Medicine. 2020;**201**(4): 378-386

[38] Kopp W, Fotter R, Steiner H, Beaufort F, Stammberger H. Aspergillosis of the paranasal sinuses. Radiology. 1985;**156**(3):715-716

[39] Dillon WP, Som PM, Fullerton GD. Hypointense MR signal in chronically inspissated sinonasal secretions. Radiology. 1990;**174**(1):73-78

[40] Agarwal R, Muthu V, Sehgal IS, Dhooria S, Prasad KT, Aggarwal AN. Allergic bronchopulmonary aspergillosis. Clinics in Chest Medicine. 2022;**43**(1): 99-125

[41] Boaventura R, Sibila O, Agusti A, Chalmers JD. Treatable traits in bronchiectasis. The European Respiratory Journal. 2018;**52**(3):1801269

[42] Patterson R, Greenberger PA, Radin RC, Roberts M. Allergic bronchopulmonary aspergillosis: Staging as an aid to management. Annals of Internal Medicine. 1982;**96**(3):286-291

[43] Agarwal R, Gupta D, Aggarwal AN, Behera D, Jindal SK. Allergic bronchopulmonary aspergillosis: Lessons from 126 patients attending a chest clinic

in North India. Chest. 2006;**130**(2): 442-448

[44] Cohen-Cymberknoh M, Blau H, Shoseyov D, Mei-Zahav M, Efrati O, Armoni S, et al. Intravenous monthly pulse methylprednisolone treatment for ABPA in patients with cystic fibrosis. Journal of Cystic Fibrosis. 2009;**8**(4): 253-257

[45] Singh Sehgal I, Agarwal R. Pulse methylprednisolone in allergic bronchopulmonary aspergillosis exacerbations. European Respiratory Review. 2014;**23**(131):149-152

[46] Agarwal R, Muthu V, Sehgal IS, Dhooria S, Prasad KT, Garg M, et al. A randomised trial of prednisolone versus prednisolone and itraconazole in acute-stage allergic bronchopulmonary aspergillosis complicating asthma. The European Respiratory Journal. 2021; **59**(4):2101787

[47] Chishimba L, Niven RM, Cooley J, Denning DW. Voriconazole and posaconazole improve asthma severity in allergic bronchopulmonary aspergillosis and severe asthma with fungal sensitization. The Journal of Asthma. 2012;**49**(4):423-433

[48] Tomomatsu K, Yasuba H, Ishiguro T, Imokawa S, Hara J, Soeda S, et al. Real-world efficacy of anti-IL-5 treatment in patients with allergic bronchopulmonary aspergillosis. Scientific Reports. 2023;**13**(1):5468

[49] Ramonell RP, Lee FE, Swenson C, Kuruvilla M. Dupilumab treatment for allergic bronchopulmonary aspergillosis: A case series. The Journal of Allergy and Clinical Immunology. In Practice. 2020; **8**(2):742-743

[50] Jin M, Douglass JA, Elborn JS, Agarwal R, Calhoun WJ, Lazarewicz S, et al. Omalizumab in allergic bronchopulmonary aspergillosis: A systematic review and meta-analysis. The Journal of Allergy and Clinical Immunology. In Practice. 2023;**11**(3): 896-905

[51] Ogata H, Sha K, Kotetsu Y, Enokizu-Ogawa A, Katahira K, Ishimatsu A, et al. Tezepelumab treatment for allergic bronchopulmonary aspergillosis. Respirology Case Reports. 2023;**11**(5): e01147

[52] Nishimura T, Okano T, Naito M, Tsuji C, Iwanaka S, Sakakura Y. Complete withdrawal of glucocorticoids after dupilumab therapy in allergic bronchopulmonary aspergillosis: A case report. World Journal of Clinical Cases. 2021;**9**(23):6922-6928

[53] Mikura S, Saraya T, Yoshida Y, Oda M, Ishida M, Honda K, et al. Successful treatment of Mepolizumab- and prednisolone-resistant allergic bronchopulmonary aspergillosis with Dupilumab. Internal Medicine. 2021; **60**(17):2839-2842

[54] Menzies-Gow A, Corren J, Bourdin A, Chupp G, Israel E, Wechsler ME, et al. Tezepelumab in adults and adolescents with severe, uncontrolled asthma. The New England Journal of Medicine. 2021;**384**: 1800-1809

Chapter 2

Post-Viral Aspergillosis

Mohammadreza Salehi, Fariba Zamani and Sadegh Khodavaisy

Abstract

Post-viral aspergillosis (PVA) is a clinical form of *Aspergillus* infection that occurs after some viral infections. *Aspergillus* is the most common respiratory fungal co-pathogen in patients with viral infections. Most cases of PVA have been reported as invasive pulmonary aspergillosis (IPA) after influenza, COVID-19, and the cytomegalovirus infection. PVA is more commonly reported in critically ill patients with viral pneumonia. Suggested risk factors for PVA include cellular immune deficiency, ARDS, pulmonary tracts and parenchyma damage, and corticosteroid therapy. New pulmonary nodules such as dense, well-circumscribed lesions with or without a halo sign, air crescent sign, or cavity, or wedge-shaped and segmental or lobar consolidation on the chest CT scan can suggest PVA. As in the treatment of invasive aspergillosis in other settings, triazoles, such as voriconazole or isavuconazole, have been suggested as the first-line treatment for PVA. It seems that the presence of PVA has significantly decreased the survival rate in patients with viral infections.

Keywords: aspergillosis, influenza, COVID-19, cytomegalovirus, viral infection

1. Introduction

Invasive pulmonary aspergillosis (IPA) is the most severe clinical form of *Aspergillus* infections and is typically seen in severely immunocompromised hosts, particularly those with hematologic malignancies undergoing chemotherapy and recipients of hematopoietic stem cell or solid organ transplantations [1, 2]. Due to the growing use of immunosuppressive agents in the treatment of many diseases and in advanced intensive care, the number of patients at risk of IPA is increasing [3]. *Aspergillus* is the most common respiratory fungal co-pathogen in patients with viral infections [4–6]. Although IPA after viral infections mainly occur in immunocompromised hosts, it has also been reported in apparently immunocompetent patients [3, 7, 8].

Viral types of pneumonia are serious health threats in the world and can occur on a seasonal, sporadic, epidemic, or even pandemic scale [9]. During the past decade, we have been faced with increasing reports of IPA in critically ill patients with viral pneumonia [10]. Viruses, such as cytomegalovirus (CMV), severe acute respiratory syndrome (SARS) virus, influenza virus, respiratory syncytial virus (RSV), parainfluenza 3 virus, and, more recently, severe respiratory syndrome coronavirus 2 (SARS-CoV-2), are among the most important causes of severe pneumonia that can cause respiratory failure and send patients to the intensive care unit (ICU) [7, 9, 11–14]. Although there are reports of the association of IPA with all severe viral pneumonia, IPA is more commonly reported in critically ill patients with influenza and COVID-19 pneumonia [14].

2. Influenza-associated pulmonary aspergillosis (IAPA)

Influenza viruses have an RNA-based genome, which lacks proofreading mechanisms, and therefore undergo constant mutations [15]. Despite medical development, Influenza is still an important virus causing serious respiratory and epidemic infections in humans and animals [16]. Influenza infections place a significant strain on health systems each year and are responsible for a large number of deaths worldwide [17]. It seems that the transmission of the influenza virus through person-to-person respiratory droplets is one of the important ways of spreading the virus and causing the disease epidemic [18]. Secondary bacterial pulmonary infections are common complications of influenza associated with a high mortality rate [19, 20]. Common bacterial pathogens causing secondary pneumonia in patients with influenza include Haemophilus influenzae, Streptococcus pyogenes, Staphylococcus aureus, and Streptococcus pneumoniae [19]. Secondary bacterial infections have been well described as complications of influenza. Pulmonary involvement with different species of *Aspergillus* also seems to be a potential complication of influenza; however, more studies are still needed to understand its different aspects [21].

The first case of IAPI was reported by JD Abbott et al. in 1952 [22]. The development of invasive fungal infections after influenza was a rare influenza complication before 2009, but the number of reported cases has been increasing since the 2009 H1N1/swine flu/influenza virus pandemic [7, 23, 24]. Almost all influenza patients with aspergillosis have had pulmonary fungal infection, but cases of tracheobronchitis and even cerebral involvement have also been reported [21].

The influenza virus infection has recently been considered as a risk factor for the development of aspergillosis in various studies [4, 5, 8]. In various studies on critically ill patients with influenza, the rate of the aspergillosis has been reported from less than 2% to more than 20% [4, 25–28]. Interestingly, to date, most reported cases of IAPA have been associated with the influenza A H1N1 subtype, but limited cases of influenza B with aspergillosis have also been presented [5, 29, 30]. The pathogenesis of IAPA is still not fully understood, but several risk factors have been mentioned [10]. Understanding the pathogenesis of IAPA requires understanding the pathogenesis of the influenza virus infection and aspergillosis and the conditions of the human host [5]. The influenza virus has been reported to cause cellular immune deficiency, alveolar epithelial damage, disruption of normal ciliary clearance in the respiratory tract, and leukopenia [31, 32]. Only few patients with the influenza infection have been reported to require hospitalization, and less than 30% of hospitalized patients have developed progressive pneumonia, but these few cases have been accompanied by a profound inflammatory response and the most severe form of acute lung injury called acute respiratory distress syndrome (ARDS) [33]. Radiological findings of the lung usually include diffuse alveolar infiltration and bilateral ground glass opacities [34]. Histopathological examination of lung parenchymal tissue in severe and fatal cases of influenza H1N1 has shown different degrees of diffuse alveolar damage with hyaline membranes and necrotizing bronchiolitis [35]. Patients with ARDS have shown the higher plasma levels of pro-inflammatory markers, such as interleukin-6, interleukin-10, and interleukin-15. than patients with the less severe disease [36]. Despite the controversial role of corticosteroids in the treatment of ARDS, it has been reported that a significant percentage of patients with ARDS secondary to the influenza infection receive corticosteroids [37, 38]. In a report of hospitalized patients with 2009 H1N1 influenza, not only patients with ARDS but also most patients without this complication received corticosteroids [34]. Classic risk factors for IPA in patients without influenza include cellular immune deficiency, chronic obstructive pulmonary disease (COPD), ICU stay, and corticosteroid therapy [39–41]. However, several reports show that critically ill patients often develop IPA even in the absence of classic risk factors [41–43]. It seems that pulmonary tracts and parenchyma damage, ARDS, immune system dysregulation, male sex, need for the prolonged ICU stay, and broad-spectrum antibiotics, and corticosteroids therapy are probably the most important risk factors for IAPA (**Figure 1**) [4, 8, 18, 44]. In a report, treatment with neuraminidase inhibitors, such as oseltamivir, was also mentioned as a possible risk factor for IAPA [45].

The European Organization for Research and Treatment of Cancer/Mycoses Study Group Education and Research Consortium (EORTC/MSGERC) has released criteria for diagnosing invasive aspergillosis in critically ill patients. The criteria include definitions for proven and probable cases although most reported patients are compatible with the definition of probable cases [21–23, 46]. Suggested criteria to define probable cases of IAPA in the ICU setting are (1) cytology, direct microscopy, and/or culture showing the presence of *Aspergillus* species in a sample of the lower respiratory tract; (2) galactomannan (GM) antigen >0.5 in plasma/serum and/or galactomannan antigen >0.8 in the bronchoalveolar lavage (BAL) specimen [46]. The diagnostic approach in most studies focuses on the BAL culture and detection of GM in serum and BAL (probable IPA) [8]. However, *Aspergillus* spp. isolated from BAL examinations in ICU patients with influenza may be overlooked as a contamination despite their potential to cause an invasive disease [3]. Although most patients under mechanical ventilation undergo bronchoscopy, the absence of a positive fungal culture does not rule out the diagnosis of IPA [41]. Although the usual radiological findings of IPA, including cavitary lesions, halo sign, or air crescent sign, have been seen only in a small number of critically ill patients, performing chest CT scan may

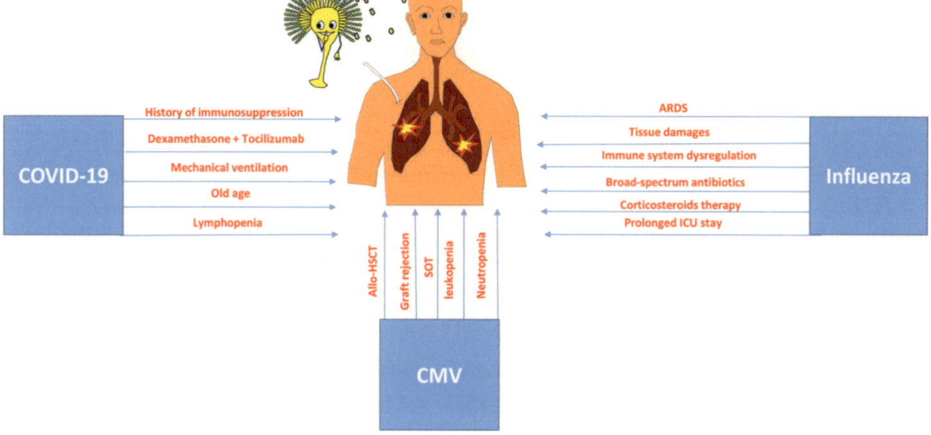

Figure 1.
Post-Viral Aspergillosis Risk Factors.

be helpful in diagnosis [21, 47]. (1-3)-β-d-glucan (BDG) is of limited value in the diagnosis of IPA; however, the combination with GM or the polymerase chain reaction (PCR) method may give this noninvasive test a more diagnostic role [48].

The mean time between the diagnosis of influenza and aspergillosis has been reported to be 6 days (range 0–32) [21]. In patients with influenza, especially in critical cases with clinical, mycological, or radiological suspicion of IPA, it is recommended to start antifungal agents (voriconazole as the treatment of choice) as soon as possible [21, 49]. In the absence of an appropriate response to treatment, therapeutic drug monitoring (TDM), evaluating resistance to azoles, and then a tissue biopsy of the suspicious lesions should be considered [50]. Complete mycological evaluations, including identification of *Aspergillus* species, are mandatory because some species are intrinsically azole resistant. Preferably, the antifungal susceptibility pattern should be performed for *Aspergillus* isolates [48].

The overall mortality rate of patients with H1N1 has been reported to be less than 0.5% [18]. It seems that IPA in critically ill patients with influenza can be associated with a poorer outcome [4]. Early reports considered the mortality rate of the IAPA at nearly 100%, but the rate has been reported between 33 and 67% in later studies, although this mortality may be higher in patients with a history of immunodeficiency [5, 8, 21, 30, 47, 51, 52]. Mortality rate in critically ill patients without influenza in ICU has been 80% and in patients with COPD has been about 95% [40, 41].

3. COVID-19 associated pulmonary aspergillosis (CAPA)

In December 2019, the first cases of pneumonia with an unknown origin were reported from Wuhan, the capital city of China's Hubei province [53]. The isolated pathogen causing this infection was identified as a novel enveloped RNA betacoronavirus, currently named severe acute respiratory syndrome coronavirus 2 (SARS-CoV-2), which is phylogenetically similar to SARS-CoV [54]. Since most of the first reported patients had contact with a southern Chinese seafood market in Wuhan, it is widely believed that COVID-19 originated from wild animals such as bats [55]. Finally, the World Health Organization (WHO) declared the 2019 coronavirus disease (COVID-19) as a public health emergency of high international concern [56]. The common transmission routes for previous coronaviruses and influenza, that is, respiratory droplets and direct contacts, are also the main ways for SARS-COV-2 transmission [57]. SARS-CoV-2 is contagious and transmissible during the incubation period and can cause numerous clusters [58].

The COVID-19 pandemic caused by SARS-CoV-2 has affected the health and life of all people in all continents and has caused a high rate of morbidity and mortality in human societies [59]. This pandemic has continued for over 3 years and is still a global threat [60]. The extensive and continuous evolution of SARS-CoV-2 has caused the emergence and spread of several variants of concern (VOCs), such as alpha (B.1.1.7), beta (B.1.351), gamma (P.1), delta (B.1.617.2), and omicron (B.1.1.529) around the world [61].

The incubation period of this infection is between 1 day and 2 weeks [62]. This viral disease usually begins with flu-like symptoms such as myalgia, fever, stuffy nose, and cough [63]. From the beginning, it was evident that ARDS is the final cause of death in many COVID-19 patients [62]. The common risk factors for the progression of the disease toward ARDS are male gender, old age, pregnant women, and the presence of underlying diseases, especially hypertension, diabetes mellitus, and cardiovascular diseases [64, 65]. The most common laboratory changes in patients with COVID-19 are lymphocytopenia, increased C-reactive protein (CRP), increased lactate dehydrogenase (LDH), and leukocytopenia [66, 67]. Several antiviral drugs and anti-inflammatory agents were examined for the treatment of COVID-19 patients, but they were unsuccessful [68–70]. At the beginning of the pandemic, corticosteroids were prescribed by many clinical teams to treat hospitalized COVID-19 patients, but after the publication of the successful results of recovery trials, the administration of dexamethasone was introduced to most therapeutic protocols to save hospitalized patients [71, 72]. Finally, in many critical patients with COVID-19, the high doses of corticosteroids (pulse therapy) were prescribed [73, 74].

Not much time had passed since the beginning of the pandemic when the possibility of invasive fungal infections such as aspergillosis, candidiasis, mucormycosis, and pneumocystosis in COVID-19 patients was raised [75]. In subsequent studies, the risk of bacterial and fungal co-infections with COVID-19 was strongly considered [76, 77]. Not only the incidence of COVID-19 co-infections reported from different medical centers is varied but also the rate of bacterial secondary infections is lower in COVID-19 patients than in patients with severe influenza [77]. The reason for the difference in the reported incidence rate of CAPA is probably the diagnostic challenges of CAPA in patients with severe COVID-19 [78]. In the absence of a comprehensive definition for CAPA, classification criteria are modified and vary widely between the studies and are often based on the mycological evidence, such as direct microscopic examination and culture or even GM testing in serum or tracheal aspirates [79, 80]. Considering the same criteria, a median prevalence of more than 20% (1.5–38%) was reported for CAPA in critical patients with COVID-19 who required invasive mechanical ventilation [80, 81]. The most common risk factors for CAPA include the history of immunosuppressive agent use, especially the combination of dexamethasone and tocilizumab, aggressive mechanical ventilation, and old age (**Figure 1**) [82].

The pathogenesis of CAPA is complex and requires the understanding of the biological and immunological processes caused by SARS-CoV-2 in the host. Similar to previous coronaviruses, SARS-CoV-2 targets and destroys epithelial cells and pneumocytes through viral protein binding to angiotensin-converting enzyme 2 (ACE2) receptors [83, 84]. Two possible mechanisms for the development of CAPA in COVID-19 patients with ARDS have been described: The first mechanism involves the release of danger-associated molecular patterns (DAMPs) by damaged cells, which act as signals intensifying the immune and inflammatory response leading to lung damage. DAMPs also develop advanced glycation end products, which integrate with Toll-like receptors (TLRs) to generate and amplify the inflammatory response in aspergillosis. Ultimately stimulation of inflammatory signals appears to increase the risk of CAPA [84]. The second mechanism can be severe lymphocytopenia, which is one of the known factors in the development of IPA in patients with hematological malignancies; however, severe lymphopenia and lymphocyte dysfunction are usually observed in patients with severe COVID-19 and probably contribute to the development of CAPA [85, 86].

The clinical features and radiological findings of CAPA are very similar to those of severe cases of COVID-19, especially cases with ARDS [78, 84]. CAPA is diagnosed in an average of 8 days (range 0–31 days) after the transfer of critically ill patients with COVID-19 to the ICU [82]. Although the radiological evidence of severe COVID-19 can be similar to that of IPA, it is recommended that a thorough work-up should be done with the observation of multiple new pulmonary nodules or lung cavities to diagnose probable CAPA as these cases are less common with COVID-19. Although some radiological features like the halo sign are typical for IPA, it is not sufficient to diagnose CAPA without mycological evidence as the halo sign is indicative of local infarction and is an intrinsic part of imaging observations of severe COVID-19 [87].

Due to the clearance of GM from the systemic blood circulation by neutrophils in non-neutropenic patients, the serum GM test might not have the necessary

CAPA definition	Host factors	Clinical features	Laboratory evidence
Possible	Need to intensive care; glucocorticoid therapy [equivalent to prednisone, 20 mg/day]; dexamethasone plus tocilizumab; history of hematological malignancy or All-HSCT or SOT or GVHD	New pulmonary infiltrate or cavity lesion on the chest CT scan that has no other causes	Presence of at least one of the following: microscopic detection of mold elements in NBL; positive NBL culture for mold; single NBL galactomannan>4.5; NBL galactomannan >1.2 twice or more; NBL galactomannan >1.2 plus another NBL mycology test positive (PCR or LFA)
Probable	Need to intensive care; glucocorticoid therapy [equivalent to prednisone, 20 mg/day]; dexamethasone plus tocilizumab; history of hematological malignancy or All-HSCT or SOT or GVHD	New pulmonary infiltrate on the chest CT scan that has no other causes as follows: dense, well-circumscribed lesions with or without a halo sign, air crescent sign, cavity, or wedge-shaped and segmental or lobar consolidation	Presence of at least one of the following: microscopic detection of fungal elements in sputum, bronchoalveolar lavage, bronchial brush, or aspirate indicating a mold; *Aspergillus* recovered by culture of bronchoalveolar lavage or bronchial brush; single serum or plasma galactomannan ≥1.0, bronchoalveolar lavage fluid galactomannan ≥1.0, single serum or plasma galactomannan ≥0.7 and bronchoalveolar lavage fluid galactomannan ≥0.8; or two or more positive *Aspergillus* PCR on plasma, serum, or whole blood; or on bronchoalveolar lavage fluid
proven	Need to intensive care; glucocorticoid therapy [equivalent to prednisone, 20 mg/day]; dexamethasone plus tocilizumab; history of hematological malignancy or All-HSCT or SOT or GVHD	New pulmonary infiltrate on the chest CT scan that has no other causes as follows: dense, well-circumscribed lesions with or without a halo sign, air crescent sign, cavity, or wedge-shaped and segmental or lobar consolidation	Presence of at least one of the following: histopathological detection of fungal hyphae showing invasive growth with associated tissue damage; *Aspergillus* recovered by culture or microscopy; or histology or PCR obtained by a sterile aspiration or biopsy from a pulmonary site

CAPA: COVID-19 associated pulmonary aspergillosis, NBL: non-bronchoscopic lavage, Allo-HSCT: allogeneic hematopoietic stem cell transplant, SOT: solid organ transplant, GVHD: Graft-versus-host disease, LFA: lateral flow assay.

Table 1.
Case definitions for patients with possible, probable, and proven CAPA.

diagnostic sensitivity for CAPA [88, 89]. Early bronchoscopy and BAL in COVID-19 patients with suspected CAPA may lead to a faster diagnosis and better management; however, bronchoscopy is rarely performed in these patients due to concerns about SARS-COV2 transmission [90]. Finally, in 2020, the diagnostic criteria for CAPA were released by the European Confederation of Medical Mycology/the International Society for Human and Animal Mycology (ECMM/ISHAM), and the definitions were described as possible, probable, and proven (**Table 1**) [87]. The implementation of noninvasive diagnostic criteria with an emphasis on the GM test, culture, PCR, and non-bronchoscopic lavage for diagnosing possible CAPA has significantly reduced the diagnosed cases of CAPA and the prevalence of CAPA to about 10% among critically ill patients with COVID-19 [82, 90, 91]. As in the treatment of invasive aspergillosis in other settings, triazoles, such as voriconazole or avuconazole, have been suggested as the first-line treatment for CAPA, and in suspected cases of resistance to azoles, liposomal amphotericin B is the main alternative [87].

It seems that the presence of CAPA has significantly decreased the survival rate in COVID-19 patients, in studies on patients with CAPA, the mortality rate has been reported to be more than 40% [78]. In a large multicenter study from French ICUs conducted on the COVID-19 patients under respiratory support with mechanical ventilation, CAPA was an independent risk factor for death, with a hazard ratio of 1.45 compared with those without the infection. In this study, the administration of triazoles, such as voriconazole and other antifungal agents, did not change the patients' outcomes [92].

4. Cytomegalovirus-associated aspergillosis (CAA)

Cytomegalovirus (CMV) is a member of the Herpesviridae family and, like the other viruses of this family, develops a persistent state after the initial acute infection, which serves as a reservoir for reactivation and subsequent infection, particularly in immunocompromised hosts [93]. CMV is transmitted through salivary secretions, sexual contact, placenta, breastfeeding, blood transfusion, and solid organ transplantation (SOT) or hematopoietic stem cell transplantation (HSCT) [94]. CMV infects many people in the world, and its primary infection is usually asymptomatic [95]. In some immunocompetent hosts, CMV can lead to a mononucleosis-like syndrome with pharyngitis, fever, myalgia, and lymphadenopathy [93]. The global prevalence of seropositive individuals for CMV has been reported to be over 80% in the general population, with the highest seroprevalence observed in the Eastern Mediterranean region of the World Health Organization (WHO) and the lowest in the European WHO region [96]. The broad cellular tropism of CMV probably contributes to the development of a diverse number of pathologies associated with the infection in different organs [97].

CMV is one of the most important pathogens that cause serious diseases in immunocompromised hosts [98]. Before the treatment of HIV/AIDS patients with antiretroviral therapy (ART), approximately 40% of people living with HIV developed diseases caused by CMV [99]. CMV is one of the most important opportunistic viruses in solid organ transplantation (SOT), causing infections and diseases, which can have adverse consequences for allograft and recipient survival, increase the patient cost, and affect the quality of life [100]. Despite the progress made in the prevention of CMV, it remains one of the main causes of morbidity and mortality after allogeneic hematopoietic stem cell transplantation (Allo-HSCT) [101]. The evidence shows that critically ill patients are at risk of developing CMV viremia or infection, with an average infection rate of 25% reported among these patients [102].

CMV seems to have immunosuppressive effects, and its infection is an independent risk factor for developing other systemic infections in SOT recipients [103–105]. Studies have shown that CMV infection aggravates the immunosuppression status including leukopenia in transplant recipients and increases not only the risk of bacterial infections but also the possibility of invasive fungal infections in these patients [106, 107]. The CMV infection has been reported to be an important risk factor for posttransplant *Pneumocystis jirovsi* pneumonia (PJP) [108, 109]. Studies have shown that neutropenia, Graft-versus-host disease (GVHD), corticosteroid therapy, lymphopenia, and CMV infection are risk factors for posttransplant aspergillosis [110, 111].

The CMV infection and IPA have been found to be important infectious diseases in transplant recipients [112]. The incidence of posttransplant IPA varies by transplant type and reporting transplant centers [113, 114]. Early IPA occurs within the first 90 days after transplantation and is more related to the hemodialysis or critical conditions of transplant recipients, while late IPA, after 90 days of transplantation, is more related to conditions of immunosuppression and the allograft rejection [115, 116]. Interestingly, in a study on lung transplant recipients with CAA, a respiratory CMV infection was seen, and the virus was previously detected in their BAL secretions [117]. CMV in transplant recipients can significantly increase the chance of CAA regardless of the transplantation type, although this may not include asymptomatic CMV viremia [112, 118]. The proposed risk factors for the development of CAA include intensified immunosuppression, higher CMV viral load, graft rejection, host genetics (polymorphisms in the toll-like receptor-4), ganciclovir-induced neutropenia, and leukopenia (**Figure 1**) [107, 119–122].

Two important points in preventing CAA in transplant patients are paying attention to the protocols for CMV prevention after transplantation and starting aspergillosis prophylaxis in the transplant recipients infected with CMV [112, 117, 123].

Timely diagnosis of CAA, treatment of both infections, attention to possible drug interactions, and reducing as much as possible the level of the patient's immunodeficiency level may reduce the risk of death [112, 117].

5. Conclusions

Post-viral aspergillosis (PVA) is a clinical form of *Aspergillus* infection that happens after some viral infections. *Aspergillus* is the most common respiratory fungal co-pathogen in patients with viral infections. Most cases of PVA have been reported as invasive pulmonary aspergillosis after influenza, COVID-19, and cytomegalovirus.

Influenza-associated pulmonary aspergillosis (IAPA): The first case of IAPI was reported in 1952. The development of an invasive fungal infections after influenza was a rare influenza complication before 2009, but the number of reported cases has been increasing since the 2009 H1N1 influenza pandemic. Almost all influenza patients with aspergillosis have had pulmonary fungal infection. The rate of IAPI has been reported from less than 2% to more than 20%. It seems that IPA in critically ill patients with influenza can be associated with a poorer outcome.

COVID-19-associated pulmonary aspergillosis (CAPA): Not much time had passed since the beginning of the pandemic when the possibility of invasive fungal infections such as aspergillosis in COVID-19 patients was raised. The clinical features and radiological findings of CAPA are very similar to those of severe cases of COVID-19, especially cases with ARDS. CAPA is diagnosed in an average of 8 days after the transfer of critically ill patients with COVID-19 to the ICU. Although the radiological evidence of severe COVID-19 can be similar to CAPA, it is recommended that a thorough work-up should be done with the observation of multiple new pulmonary nodules or lung cavities to diagnose probable CAPA as these cases are less common with COVID-19. It seems that the presence of CAPA has significantly decreased the survival rate in COVID-19 patients.

Cytomegalovirus-associated aspergillosis (CAA): CMV seems to have immunosuppressive effects, and its infection is an independent risk factor for developing other systemic infections such as fungal infections. CMV in transplant recipients can significantly increase the chance of CAA regardless of the transplantation type, although this may not include asymptomatic CMV viremia. The proposed risk factors for the development of CAA include intensified immunosuppression, higher CMV viral load, graft rejection, host genetics (polymorphisms in the toll-like receptor-4), ganciclovir-induced neutropenia, and leukopenia. Timely diagnosis of CAA, treatment of both infections, attention to possible drug interactions, and reducing as much as possible the level of the patient's immunodeficiency level may reduce the risk of death.

Acknowledgements

We would like to thank the staff of research center for antibiotic stewardship and antimicrobial resistance of Tehran University of Medical Sciences, Tehran, Iran.

Conflict of interest

The authors declare no conflict of interest.

Author details

Mohammadreza Salehi[1]*, Fariba Zamani[2] and Sadegh Khodavaisy[3]

1 Research Center for Antibiotic Stewardship and Anti-microbial Resistance, Department of Infectious Diseases, Tehran University of Medical Sciences, Tehran, Iran

2 Research Center for Antibiotic Stewardship and Anti-microbial Resistance, Tehran University of Medical Sciences, Tehran, Iran

3 Department of Medical Parasitology and Mycology, School of Public Health, Tehran University of Medical Sciences, Tehran, Iran

*Address all correspondence to: salehi.mohamad3@gmail.com; mr-salehi@sina.tums.ac.ir

References

[1] Rahi MS, Jindal V, Pednekar P, Parekh J, Gunasekaran K, Sharma S, et al. Fungal infections in hematopoietic stem-cell transplant patients: A review of epidemiology, diagnosis, and management. Therapeutic Advances in Infectious Disease. 2021;**8**:20499361211039050

[2] Mühlemann K, Wenger C, Zenhäusern R, Täuber MG. Risk factors for invasive aspergillosis in neutropenic patients with hematologic malignancies. Leukemia. 2005;**19**(4):545-550

[3] Lat A, Bhadelia N, Miko B, Furuya EY, Thompson GR 3rd. Invasive aspergillosis after pandemic (H1N1) 2009. Emerging Infectious Diseases. 2010;**16**(6):971-973

[4] Waldeck F, Boroli F, Suh N, Wendel Garcia PD, Flury D, Notter J, et al. Influenza-associated aspergillosis in critically-ill patients—A retrospective bicentric cohort study. European Journal of Clinical Microbiology & Infectious Diseases. 2020;**39**:1915-1923

[5] Van De Veerdonk FL, Kolwijck E, Lestrade PP, Hodiamont CJ, Rijnders BJ, Van Paassen J, et al. Influenza-associated aspergillosis in critically ill patients. American Journal of Respiratory and Critical Care Medicine. 2017;**196**(4):524-527

[6] Salehi M, Khajavirad N, Seifi A, Salahshour F, Jahanbin B, Kazemizadeh H, et al. Proven Aspergillus flavus pulmonary aspergillosis in a COVID-19 patient: A case report and review of the literature. Mycoses. 2021;**64**(8):809-816

[7] Garcia-Vidal C, Barba P, Arnan M, Moreno A, Ruiz-Camps I, Gudiol C, et al. Invasive aspergillosis complicating pandemic influenza A (H1N1) infection in severely immunocompromised patients. Clinical Infectious Diseases. 2011;**53**(6):e16-e19

[8] Wauters J, Baar I, Meersseman P, Meersseman W, Dams K, De Paep R, et al. Invasive pulmonary aspergillosis is a frequent complication of critically ill H1N1 patients: A retrospective study. Intensive Care Medicine. 2012;**38**:1761-1768

[9] Ruuskanen O, Lahti E, Jennings LC, Murdoch DR. Viral pneumonia. Lancet. 2011;**377**(9773):1264-1275

[10] Dewi IM, Janssen NA, Rosati D, Bruno M, Netea MG, Brüggemann RJ, et al. Invasive pulmonary aspergillosis associated with viral pneumonitis. Current Opinion in Microbiology. 2021;**62**:21-27

[11] Salomon N, Perlman DC. Cytomegalovirus pneumonia. Seminars in Respiratory infections. 1999;**14**(4):353-358

[12] Short KR, Kroeze E, Fouchier RAM, Kuiken T. Pathogenesis of influenza-induced acute respiratory distress syndrome. The Lancet Infectious Diseases. 2014;**14**(1):57-69

[13] Attaway AH, Scheraga RG, Bhimraj A, Biehl M, Hatipoğlu U. Severe covid-19 pneumonia: Pathogenesis and clinical management. BMJ. 2021;**372**:n436

[14] Cillóniz C, Pericàs JM, Rojas JR, Torres A. Severe infections due to respiratory viruses. Seminars in Respiratory and Critical Care Medicine. 2022;**43**(1):60-74

[15] Webster RG, Govorkova EA. Continuing challenges in influenza.

Annals of the New York Academy of Sciences. 2014;**1323**(1):115-139

[16] Palese P. Influenza: Old and new threats. Nature Medicine. 2004;**10**(Suppl. 12):S82-SS7

[17] Iuliano AD, Roguski KM, Chang HH, Muscatello DJ, Palekar R, Tempia S, et al. Estimates of global seasonal influenza-associated respiratory mortality: A modelling study. Lancet. 2018;**391**(10127):1285-1300

[18] Bautista E, Chotpitayasunondh T, Gao Z, Harper SA, Shaw M, Uyeki TM, et al. Clinical aspects of pandemic 2009 influenza A (H1N1) virus infection. The New England Journal of Medicine. 2010;**362**(18):1708-1719

[19] McCullers JA. The co-pathogenesis of influenza viruses with bacteria in the lung. Nature Reviews Microbiology. 2014;**12**(4):252-262

[20] Rice TW, Rubinson L, Uyeki TM, Vaughn FL, John BB, Miller RR III, et al. Critical illness from 2009 pandemic influenza A (H1N1) virus and bacterial co-infection in the United States. Critical Care Medicine. 2012;**40**(5):1487

[21] Crum-Cianflone NF, editor. Invasive aspergillosis associated with severe influenza infections. In: Open Forum Infectious Diseases. Oxford University Press; 2016

[22] Abbott J, Fernando H, Gurling K, MeadeB.Pulmonaryaspergillosisfollowing post-influenzal bronchopneumonia treated with antibiotics. British Medical Journal. 1952;**1**(4757):523

[23] Adalja AA, Sappington PL, Harris SP, Rimmele T, Kreit JW, Kellum JA, et al. Isolation of Aspergillus in three 2009 H1N1 influenza patients. Influenza and Other Respiratory Viruses. 2011;**5**(4):225-229

[24] Lat A, Bhadelia N, Miko B, Furuya EY, Thompson GR. Invasive aspergillosis after pandemic (H1N1) 2009. Infectious diseases. 2010;**16**(6):971-973

[25] Martin-Loeches I, Schultz J, Vincent J-L, Alvarez-Lerma F, Bos LD, Solé-Violán J, et al. Increased incidence of co-infection in critically ill patients with influenza. Intensive Care Medicine. 2017;**43**:48-58

[26] Martin-Loeches I, Lisboa T, Rhodes A, Moreno RP, Silva E, Sprung C, et al. Use of early corticosteroid therapy on ICU admission in patients affected by severe pandemic (H1N1) v influenza A infection. Intensive Care Medicine. 2011;**37**:272-283

[27] Passouant O, Mateu P, Commandini M, Brenkle K, Just B, editors. Pulmonary Aspergillosis in Non-immunocompromised Patient with Acute Respiratory Distress Syndrome During A (H1N1) Infection. Annales Francaises D'anesthesie et de Reanimation; 2011

[28] Kim SH, Hong SB, Yun SC, Choi WI, Ahn JJ, Lee YJ, et al. Corticosteroid treatment in critically ill patients with pandemic influenza A/H1N1 2009 infection: Analytic strategy using propensity scores. American Journal of Respiratory and Critical Care Medicine. 2011;**183**(9):1207-1214

[29] Nulens EF, Bourgeois MJ, Reynders MB. Post-influenza aspergillosis, do not underestimate influenza B. Infection and Drug Resistance. 2017;**10**:61-67

[30] Schauwvlieghe AF, Rijnders BJ, Philips N, Verwijs R, Vanderbeke L, Van Tienen C, et al. Invasive aspergillosis

in patients admitted to the intensive care unit with severe influenza: A retrospective cohort study. The Lancet Respiratory Medicine. 2018;**6**(10):782-792

[31] Lewis DE, Gilbert BE, Knight V. Influenza virus infection induces functional alterations in peripheral blood lymphocytes. Journal of Immunology. 1986;**137**(12):3777-3781

[32] Sedger LM, Hou S, Osvath SR, Glaccum MB, Peschon JJ, van Rooijen N, et al. Bone marrow B cell apoptosis during in vivo influenza virus infection requires TNF-alpha and lymphotoxin-alpha. Journal of Immunology. 2002;**169**(11):6193-6201

[33] Ramsey C, Kumar A. H1N1: Viral pneumonia as a cause of acute respiratory distress syndrome. Current Opinion in Critical Care. 2011;**17**(1):64-71

[34] Jain S, Kamimoto L, Bramley AM, Schmitz AM, Benoit SR, Louie J, et al. Hospitalized patients with 2009 H1N1 influenza in the United States, April–June 2009. New England Journal of Medicine. 2009;**361**(20):1935-1944

[35] Shieh W-J, Blau DM, Denison AM, DeLeon-Carnes M, Adem P, Bhatnagar J, et al. 2009 pandemic influenza A (H1N1): Pathology and pathogenesis of 100 fatal cases in the United States. The American Journal of Pathology. 2010;**177**(1):166-175

[36] To KK, Hung IF, Li IW, Lee K-L, Koo C-K, Yan W-W, et al. Delayed clearance of viral load and marked cytokine activation in severe cases of pandemic H1N1 2009 influenza virus infection. Clinical Infectious Diseases. 2010;**50**(6):850-859

[37] Domínguez-Cherit G, Lapinsky SE, Macias AE, Pinto R, Espinosa-Perez L, de la Torre A, et al. Critically Ill patients with 2009 influenza A(H1N1) in Mexico. Journal of the American Medical Association. 2009;**302**(17):1880-1887

[38] Webb SA, Pettilä V, Seppelt I, Bellomo R, Bailey M, Cooper DJ, et al. Critical care services and 2009 H1N1 influenza in Australia and New Zealand. The New England Journal of Medicine. 2009;**361**(20):1925-1934

[39] Allam MF, Castillo AS, Díaz-Molina C, Navajas RF-C. Invasive pulmonary aspergillosis: Identification of risk factors. Scandinavian Journal of Infectious Diseases. 2002;**34**(11):819-822

[40] Ader F, Nseir S, Le Berre R, Leroy S, Tillie-Leblond I, Marquette C, et al. Invasive pulmonary aspergillosis in chronic obstructive pulmonary disease: An emerging fungal pathogen. Clinical Microbiology and Infection. 2005;**11**(6):427-429

[41] Meersseman W, Vandecasteele SJ, Wilmer A, Verbeken E, Peetermans WE, Van Wijngaerden E. Invasive aspergillosis in critically ill patients without malignancy. American Journal of Respiratory and Critical Care Medicine. 2004;**170**(6):621-625

[42] Vandewoude KH, Blot SI, Depuydt P, Benoit D, Temmerman W, Colardyn F, et al. Clinical relevance of Aspergillus isolation from respiratory tract samples in critically ill patients. Critical Care. 2006;**10**(1):1-10

[43] Meersseman W, Lagrou K, Maertens J, Wijngaerden EV. Invasive aspergillosis in the intensive care unit. Clinical Infectious Diseases. 2007;**45**(2):205-216

[44] Ahmadikia K, Hashemi SJ, Khodavaisy S, Getso MI, Alijani N, Badali H, et al. The double-edged sword

of systemic corticosteroid therapy in viral pneumonia: A case report and comparative review of influenza-associated mucormycosis versus COVID-19 associated mucormycosis. Mycoses. 2021;**64**(8):798-808

[45] Verweij PE, Rijnders BJ, Brüggemann RJ, Azoulay E, Bassetti M, Blot S, et al. Review of influenza-associated pulmonary aspergillosis in ICU patients and proposal for a case definition: An expert opinion. Intensive Care Medicine. 2020;**46**(8):1524-1535

[46] Bassetti M, Azoulay E, Kullberg B-J, Ruhnke M, Shoham S, Vazquez J, et al. EORTC/MSGERC definitions of invasive fungal diseases: Summary of activities of the Intensive Care Unit Working Group. Clinical Infectious Diseases. 2021;**72**(Supplement_2):S121-S1S7

[47] Ku Y-H, Chan K-S, Yang C-C, Tan C-K, Chuang Y-C, Yu W-L. Higher mortality of severe influenza patients with probable aspergillosis than those with and without other coinfections. Journal of the Formosan Medical Association. 2017;**116**(9):660-670

[48] Ullmann AJ, Aguado JM, Arikan-Akdagli S, Denning DW, Groll AH, Lagrou K, et al. Diagnosis and management of Aspergillus diseases: Executive summary of the 2017 ESCMID-ECMM-ERS guideline. Clinical Microbiology and Infection. 2018;**24**:e1-e38

[49] Vanderbeke L, Spriet I, Breynaert C, Rijnders BJ, Verweij PE, Wauters J. Invasive pulmonary aspergillosis complicating severe influenza: Epidemiology, diagnosis and treatment. Current Opinion in Infectious Diseases. 2018;**31**(6):471-480

[50] Koehler P, Bassetti M, Kochanek M, Shimabukuro-Vornhagen A, Cornely OA. Intensive care management of influenza-associated pulmonary aspergillosis. Clinical Microbiology and Infection. 2019;**25**(12):1501-1509

[51] Lewis M, Kallenbach J, Ruff P, Zaltzman M, Abramowitz J, Zwi S. Invasive pulmonary aspergillosis complicating influenza A pneumonia in a previously healthy patient. Chest. 1985;**87**(5):691-693

[52] Boots R, Paterson D, Allworth A, Faoagali J. Successful treatment of post-influenza pseudomembranous necrotising bronchial aspergillosis with liposomal amphotericin, inhaled amphotericin B, gamma interferon and GM-CSF. Thorax. 1999;**54**(11):1047-1049

[53] Huang C, Wang Y, Li X, Ren L, Zhao J, Hu Y, et al. Clinical features of patients infected with 2019 novel coronavirus in Wuhan, China. The Lancet. 2020;**395**(10223):497-506

[54] Guan W-j, Ni Z-y, Hu Y, Liang W-h, Ou C-q, He J-x, et al. Clinical characteristics of coronavirus disease 2019 in China. New England Journal of Medicine. 2020;**382**(18):1708-1720

[55] Zhou P, Yang X-L, Wang X-G, Hu B, Zhang L, Zhang W, et al. A pneumonia outbreak associated with a new coronavirus of probable bat origin. Nature. 2020;**579**(7798):270-273

[56] Organization WH. World Health Organization Coronavirus Disease (COVID-19) Dashboard. Geneva: World Health Organization; 2020

[57] Rothe C, Schunk M, Sothmann P, Bretzel G, Froeschl G, Wallrauch C, et al. Transmission of 2019-nCoV infection from an asymptomatic contact in Germany. New England Journal of Medicine. 2020;**382**(10):970-971

[58] Huang X, Wei F, Hu L, Wen L, Chen K. Epidemiology and clinical characteristics of COVID-19. Archives of Iranian Medicine. 2020;**23**(4):268-271

[59] COVID-19 Excess Mortality Collaborators. Estimating excess mortality due to the COVID-19 pandemic: A systematic analysis of COVID-19- related mortality, 2020-21. Lancet. 2022;**399**(10334):1513-1536

[60] Pan Y, Wang L, Feng Z, Xu H, Li F, Shen Y, et al. Characterisation of SARS-CoV-2 variants in Beijing during 2022: An epidemiological and phylogenetic analysis. The Lancet. 2023;**401**(10377):664-672

[61] WHO. Tracking SARS-CoV-2 variants. 2022

[62] Wang W, Tang J, Wei F. Updated understanding of the outbreak of 2019 novel coronavirus (2019-nCoV) in Wuhan, China. Journal of Medical Virology. 2020;**92**(4):441-447

[63] Yang X, Yu Y, Xu J, Shu H, Liu H, Wu Y, et al. Clinical course and outcomes of critically ill patients with SARS-CoV-2 pneumonia in Wuhan, China: A single-centered, retrospective, observational study. The Lancet Respiratory Medicine. 2020;**8**(5):475-481

[64] Allameh SF, Nemati S, Ghalehtaki R, Mohammadnejad E, Aghili SM, Khajavirad N, et al. Clinical characteristics and outcomes of 905 COVID-19 patients admitted to Imam Khomeini hospital complex in the capital city of Tehran, Iran. Archives of Iranian Medicine. 2020;**23**(11):766-775

[65] Chen N, Zhou M, Dong X, Qu J, Gong F, Han Y, et al. Epidemiological and clinical characteristics of 99 cases of 2019 novel coronavirus pneumonia in Wuhan, China: A descriptive study. The Lancet. 2020;**395**(10223):507-513

[66] Lq L, Huang T, Yq W, Zp W, Liang Y, Tb H, et al. COVID-19 patients' clinical characteristics, discharge rate, and fatality rate of meta-analysis. Journal of Medical Virology. 2020;**92**(6):577-583

[67] Ghazanfari T, Salehi MR, Namaki S, Arabkheradmand J, Rostamian A, Rajabnia Chenary M, et al. Interpretation of Hematological, Biochemical, and Immunological Findings of COVID-19 Disease: Biomarkers Associated with Severity and Mortality. Iranian Journal of Allergy, Asthma, and Immunology. 2021;**20**(1):46-66

[68] Salehi M, Edalatifard M, Taslimi R, Ghiasvand F, Khajavirad N, Mirfazaelian H. Fighting COVID-19: What are the available options? Frontiers in Mergency Medicine. 2020;**4**(2s):e65

[69] Hosseini H, Sadeghi A, Tabarsi P, Etemadimanesh A, Darazam IA, Aghdami N, et al. Another step toward final call on Remdesivir efficacy as a treatment for hospitalized COVID-19 patients: A multicenter open-label trial. MedRxiv. 2021;**2021**(08):13.21261992

[70] Salehi M, Barkhori Mehni M, Akbarian M, Fattah Ghazi S, Khajavi Rad N, Moradi Moghaddam O, et al. The outcome of using intravenous immunoglobulin (IVIG) in critically ill COVID-19 patients': A retrospective, multi-centric cohort study. European Journal of Medical Research. 2022;**27**(1):18

[71] Abbasian L, Toroghi N, Rahmani H, Khalili H, Hasannezhad M, Ghiasvand F, et al. Comparing efficacy and safety of tocilizumab and methylprednisolone in the treatment of patients with severe COVID-19. International Immunopharmacology. 2022;**107**:108689

[72] Group RC. Dexamethasone in hospitalized patients with Covid-19.

New England Journal of Medicine. 2021;**384**(8):693-704

[73] Edalatifard M, Akhtari M, Salehi M, Naderi Z, Jamshidi A, Mostafaei S, et al. Intravenous methylprednisolone pulse as a treatment for hospitalised severe COVID-19 patients: Results from a randomised controlled clinical trial. European Respiratory Journal. 2020;**56**(6):2002808

[74] Salvarani C, Massari M, Costantini M, Merlo DF, Mariani GL, Viale P, et al. Intravenous methylprednisolone pulses in hospitalised patients with severe COVID-19 pneumonia: A double-blind, randomised, placebo-controlled trial. European Respiratory Journal. 2022;**60**(4):2200025

[75] Salehi M, Ahmadikia K, Badali H, Khodavaisy S. Opportunistic fungal infections in the epidemic area of COVID-19: A clinical and diagnostic perspective from Iran. Mycopathologia. 2020;**185**(4):607-611

[76] Zhu X, Ge Y, Wu T, Zhao K, Chen Y, Wu B, et al. Co-infection with respiratory pathogens among COVID-2019 cases. Virus Research. 2020;**285**:198005

[77] Rawson TM, Moore LSP, Zhu N, Ranganathan N, Skolimowska K, Gilchrist M, et al. Bacterial and fungal coinfection in individuals with coronavirus: A rapid review to support COVID-19 antimicrobial prescribing. Clinical Infectious Diseases. 2020;**71**(9):2459-2468

[78] Hoenigl M, Seidel D, Sprute R, Cunha C, Oliverio M, Goldman GH, et al. COVID-19-associated fungal infections. Nature Microbiology. 2022;**7**(8):1127-1140

[79] Rutsaert L, Steinfort N, Van Hunsel T, Bomans P, Naesens R, Mertes H, et al. COVID-19-associated invasive pulmonary aspergillosis. Annals of Intensive Care. 2020;**10**(1):71

[80] Lamoth F, Glampedakis E, Boillat-Blanco N, Oddo M, Pagani J-L. Incidence of invasive pulmonary aspergillosis among critically ill COVID-19 patients. Clinical Microbiology and Infection. 2020;**26**(12):1706-1708

[81] Alanio A, Dellière S, Fodil S, Bretagne S, Mégarbane B. Prevalence of putative invasive pulmonary aspergillosis in critically ill patients with COVID-19. The Lancet Respiratory Medicine. 2020;**8**(6):e48-ee9

[82] Prattes J, Wauters J, Giacobbe DR, Salmanton-García J, Maertens J, Bourgeois M, et al. Risk factors and outcome of pulmonary aspergillosis in critically ill coronavirus disease 2019 patients - A multinational observational study by the European Confederation of Medical Mycology. Clinical Microbiology and Infection. 2022;**28**(4):580-587

[83] Kuba K, Imai Y, Rao S, Gao H, Guo F, Guan B, et al. A crucial role of angiotensin converting enzyme 2 (ACE2) in SARS coronavirus-induced lung injury. Nature Medicine. 2005;**11**(8):875-879

[84] Arastehfar A, Carvalho A, van de Veerdonk FL, Jenks JD, Koehler P, Krause R, et al. COVID-19 associated pulmonary aspergillosis (CAPA)-from immunology to treatment. Journal of Fungi (Basel). 2020;**6**(2):91

[85] Stanzani M, Vianelli N, Cavo M, Kontoyiannis DP, Lewis RE. Development and internal validation of a model for predicting 60-day risk of invasive mould disease in patients with haematological malignancies. Journal of Infection. 2019;**78**(6):484-490

[86] Kvedaraite E, Hertwig L, Sinha I, Ponzetta A, Hed Myrberg I,

Lourda M, et al. Major alterations in the mononuclear phagocyte landscape associated with COVID-19 severity. Proceedings of the National Academy of Sciences. 2021;**118**(6):e2018587118

[87] Koehler P, Bassetti M, Chakrabarti A, Chen SC, Colombo AL, Hoenigl M, et al. Defining and managing COVID-19-associated pulmonary aspergillosis: The 2020 ECMM/ISHAM consensus criteria for research and clinical guidance. The Lancet Infectious Diseases. 2021;**21**(6):e149-ee62

[88] Arastehfar A, Carvalho A, Houbraken J, Lombardi L, Garcia-Rubio R, Jenks J, et al. Aspergillus fumigatus and aspergillosis: From basics to clinics. Studies in Mycology. 2021;**100**(1):100115

[89] Petraitiene R, Petraitis V, Bacher JD, Finkelman MA, Walsh TJ. Effects of host response and antifungal therapy on serum and BAL levels of galactomannan and (1→ 3)-β-D-glucan in experimental invasive pulmonary aspergillosis. Medical Mycology. 2015;**53**(6):558-568

[90] Fekkar A, Neofytos D, Nguyen M-H, Clancy CJ, Kontoyiannis DP, Lamoth F. COVID-19-associated pulmonary aspergillosis (CAPA): How big a problem is it? Clinical Microbiology and Infection. 2021;**27**(9):1376-1378

[91] Janssen NA, Nyga R, Vanderbeke L, Jacobs C, Ergün M, Buil JB, et al. Multinational observational cohort study of COVID-19–associated pulmonary Aspergillosis. Emerging Infectious Diseases. 2021;**27**(11):2892

[92] Gangneux J-P, Dannaoui E, Fekkar A, Luyt C-E, Botterel F, De Prost N, et al. Fungal infections in mechanically ventilated patients with COVID-19 during the first wave: The French multicentre MYCOVID study. The Lancet Respiratory Medicine. 2022;**10**(2):180-190

[93] Gandhi MK, Khanna R. Human cytomegalovirus: Clinical aspects, immune regulation, and emerging treatments. The Lancet Infectious Diseases. 2004;**4**(12):725-738

[94] Sia IG, Patel R. New strategies for prevention and therapy of cytomegalovirus infection and disease in solid-organ transplant recipients. Clinical Microbiology Reviews. 2000;**13**(1):83-121

[95] Griffiths P, Baraniak I, Reeves M. The pathogenesis of human cytomegalovirus. The Journal of Pathology. 2015;**235**(2):288-297

[96] Zuhair M, Smit GSA, Wallis G, Jabbar F, Smith C, Devleesschauwer B, et al. Estimation of the worldwide seroprevalence of cytomegalovirus: A systematic review and meta-analysis. Reviews in Medical Virology. 2019;**29**(3):e2034

[97] Cha TA, Tom E, Kemble GW, Duke GM, Mocarski ES, Spaete RR. Human cytomegalovirus clinical isolates carry at least 19 genes not found in laboratory strains. Journal of Virology. 1996;**70**(1):78-83

[98] Hodinka RL. Human cytomegalovirus. Manual of Clinical Microbiology. 2015:1718-1737

[99] Steininger C, Puchhammer-Stöckl E, Popow-Kraupp T. Cytomegalovirus disease in the era of highly active antiretroviral therapy (HAART). Journal of Clinical Virology. 2006;**37**(1):1-9

[100] Grossi PA, Kamar N, Saliba F, Baldanti F, Aguado JM, Gottlieb J, et al. Cytomegalovirus management in solid organ transplant recipients:

A pre-covid-19 survey from the working group of the european society for organ transplantation. Transplant International. 2022;**35**:10332

[101] Jakharia N, Howard D, Riedel DJ. CMV infection in hematopoietic stem cell transplantation: Prevention and treatment strategies. Current Treatment Options in Infectious Diseases. 2021;**13**(3):123-140

[102] Kuo C-W, Wang S-Y, Tsai H-P, Su P-L, Cia C-T, Lai C-H, et al. Invasive pulmonary aspergillosis is associated with cytomegalovirus viremia in critically ill patients - A retrospective cohort study. Journal of Microbiology, Immunology and Infection. 2022;**55**(2):291-299

[103] Collins LA, Samore MH, Roberts MS, Luzzati R, Jenkins RL, Lewis WD, et al. Risk factors for invasive fungal infections complicating orthotopic liver transplantation. The Journal of Infectious Diseases. 1994;**170**(3):644-652

[104] Kotton CN, Kamar N. New insights on CMV management in solid organ transplant patients: Prevention, treatment, and management of resistant/refractory disease. Infectious Diseases and Therapy. 2023;**12**(2):333-342

[105] Boeckh M, Nichols WG. Immunosuppressive effects of beta-herpesviruses. Herpes: The Journal of the IHMF. 2003;**10**(1):12-16

[106] Nichols WG, Corey L, Gooley T, Davis C, Boeckh M. High risk of death due to bacterial and fungal infection among cytomegalovirus (CMV)—seronegative recipients of stem cell transplants from seropositive donors: evidence for indirect effects of primary CMV infection. The Journal of Infectious Diseases. 2002;**185**(3):273-282

[107] George MJ, Snydman DR, Werner BG, Griffith J, Falagas ME, Dougherty NN, et al. The independent role of cytomegalovirus as a risk factor for invasive fungal disease in orthotopic liver transplant recipients. The American Journal of Medicine. 1997;**103**(2):106-113

[108] Fishman JA, Gans H, Practice AIDCo. Pneumocystis jiroveci in solid organ transplantation: Guidelines from the American Society of Transplantation Infectious Diseases Community of Practice. Clinical Transplantation. 2019;**33**(9):e13587

[109] Hosseini-Moghaddam SM, Krishnan RJ, Guo H, Kumar D. Cytomegalovirus infection and graft rejection as risk factors for pneumocystis pneumonia in solid organ transplant recipients: A systematic review and meta-analysis. Clinical Transplantation. 2018;**32**(8):e13339

[110] Marr KA, Carter RA, Boeckh M, Martin P, Corey L. Invasive aspergillosis in allogeneic stem cell transplant recipients: Changes in epidemiology and risk factors. Blood, The Journal of the American Society of Hematology. 2002;**100**(13):4358-4366

[111] Wald A, Leisenring W, van Burik J-A, Bowden RA. Epidemiology of Aspergillus infections in a large cohort of patients undergoing bone marrow transplantation. The Journal of Infectious Diseases. 1997;**175**(6):1459-1466

[112] Chuleerarux N, Thongkam A, Manothummetha K, Nematollahi S, Dioverti-Prono V, Torvorapanit P, et al. Does posttransplant cytomegalovirus increase the risk of invasive aspergillosis in solid organ transplant recipients? A systematic review and meta-analysis. Journal of Fungi (Basel). 2021;7(5):327

[113] Kusne S, Furukawa H, Abu-Elmagd K, Irish W, Rakela J, Fung J,

et al., editors. Infectious complications after small bowel transplantation in adults: an update. In: Transplantation Proceedings. NIH Public Access; 1996

[114] Fortún J, Martín-Dávila P, Moreno S, De Vicente E, Nuño J, Candelas A, et al. Risk factors for invasive aspergillosis in liver transplant recipients. Liver Transplantation. 2002;**8**(11):1065-1070

[115] Gavalda J, Len O, San Juan R, Aguado J, Fortun J, Lumbreras C, et al. Risk factors for invasive aspergillosis in solid-organ transplant recipients: A case-control study. Clinical Infectious Diseases. 2005;**41**(1):52-59

[116] Heylen L, Maertens J, Naesens M, Van Wijngaerden E, Lagrou K, Bammens B, et al. Invasive aspergillosis after kidney transplant: Case-control study. Clinical Infectious Diseases. 2015;**60**(10):1505-1511

[117] Wulff SM, Perch M, Helweg-Larsen J, Bredahl P, Arendrup MC, Lundgren J, et al. Associations between invasive aspergillosis and cytomegalovirus in lung transplant recipients: A nationwide cohort study. APMIS. 2023. Online ahead of print. DOI: 10.1111/apm.13317

[118] Giménez E, Solano C, Nieto J, Remigia MJ, Clari MÁ, Costa E, et al. An investigation on the relationship between the occurrence of CMV DNAemia and the development of invasive aspergillosis in the allogeneic stem cell transplantation setting. Journal of Medical Virology. 2014;**86**(4):568-575

[119] Atalla A, Garnica M, Maiolino A, Nucci M. Risk factors for invasive mold diseases in allogeneic hematopoietic cell transplant recipients. Transplant Infectious Disease. 2015;**17**(1):7-13

[120] Laursen A, Mogensen S, Andersen H, Andersen P, Ellermann-Eriksen S. The impact of CMV on the respiratory burst of macrophages in response to Pneumocystis carinii. Clinical & Experimental Immunology. 2001;**123**(2):239-246

[121] Bochud P-Y, Chien JW, Marr KA, Leisenring WM, Upton A, Janer M, et al. Toll-like receptor 4 polymorphisms and aspergillosis in stem-cell transplantation. New England Journal of Medicine. 2008;**359**(17):1766-1777

[122] El Helou G, Razonable RR. Safety considerations with current and emerging antiviral therapies for cytomegalovirus infection in transplantation. Expert Opinion on Drug Safety. 2019;**18**(11):1017-1030

[123] Ekenberg C, da Cunha-Bang C, Lodding IP, Sørensen SS, Sengeløv H, Perch M, et al. Evaluation of an electronic, patient-focused management system aimed at preventing cytomegalovirus disease following solid organ transplantation. Transplant Infectious Disease. 2020;**22**(2):e13252

Chapter 3

Aspergillus and Aspergillosis in People with Chronic Diseases

Bismark Dabuo, Nunekpeku Xorlali, Ndego Timothy Amoliga, Zyaara Kono Atibodu, Precious Mavis Newman, Alhassan Mohammed, Raymond Adongsakiya Ali and Abubakari Abudu

Abstract

Numerous human diseases are caused by *Aspergillus* species. Mold infections can be more severe in people with weakened immune systems and chronic illnesses. People with underlying chronic conditions are more likely to contract an *Aspergillus* infection than immunocompromised patients, who are more likely to develop an invasive infection with these opportunistic molds. These disorders include *Aspergillus* bronchitis, allergic bronchopulmonary aspergillosis, diabetes, cystic fibrosis, severe asthma with fungal sensitivity, and other inflammatory and allergic conditions. The impact of *Aspergillus* infections in patients with selected chronic infections and the treatment of these infections are discussed in this review along with the most recent research on these topics.

Keywords: cystic fibrosis, chronic disease, immunocompromised, *Aspergillus*, hypersensitive syndrome

1. Introduction

According to Merad et al. [1], *Aspergillus* is a form of fungi that is frequently found in a variety of environmental niches, including soil, decaying plant matter, and indoor air. *Aspergillus* recolonizes indoor areas with high humidity levels, such as air conditioning units and wet carpets [2]. Even though the genus *Aspergillus* contains more than 300 species, *Aspergillus fumigatus*, *Aspergillus flavus*, *Aspergillus niger*, and *Aspergillus terreus* have a disproportionate amount of pathogenicity. The most well-known and often found pathogenic species is *Aspergillus fumigatus*, followed by *A. flavus*; however, *A. niger* and *A. terreus* also exhibit virulence. As a saprophyte organism, *Aspergillus* is capable of colonizing the mucus masses and residual cavities of individuals with chronic obstructive pulmonary disease (COPD) [3].

Although *Aspergillus* infection often does not harm healthy people, people with chronic conditions are more likely to contract the infection, which could have serious negative effects on their health. Chronic conditions including asthma, chronic obstructive pulmonary disease (COPD), cystic fibrosis, and diabetes impair the immune system and make a person more susceptible to infections. *Aspergillus* bronchitis, *Aspergillus* pneumonia, and allergic bronchopulmonary aspergillosis (ABPA) are only a few of the respiratory conditions that can develop after inhaling *Aspergillus* spores [4]. An important consequence of *Aspergillus* infection in people with chronic medical problems is an increased incidence of morbidity and mortality. These people with chronic illnesses who contract an *Aspergillus* infection are more likely to endure worsened symptoms, extended hospital stays, and complications such as the potentially fatal invasive aspergillosis. Additionally, *Aspergillus* infection may worsen preexisting chronic disorders, creating new difficulties in treating them [5]. *Aspergillus* colonization of the airways is linked to an increased incidence of disease exacerbations and a reduction in lung function, making *Aspergillus* infection a serious hazard to people with cystic fibrosis. *Aspergillus* infection can sporadically result in deadly invasive pulmonary aspergillosis (IPA) in cystic fibrosis (CF) patients. *Aspergillus* infections can harm a person's health if they already have a chronic illness [6]. To reduce their risk of infection, people with chronic illnesses must take certain precautions. These precautions include avoiding settings with a lot of mold and fungi, practicing excellent hygiene, and seeking immediate medical attention if they feel symptoms of an infection. The clinical progression of chronic disorders can be significantly influenced by *Aspergillus* and aspergillosis. People who already have respiratory conditions like asthma, COPD, or cystic fibrosis run the risk of having their symptoms worsened by *Aspergillus*, which could result in progressive lung damage. Long-term *Aspergillus* exposure, especially in susceptible people, may be a factor in the emergence of chronic lung diseases. The therapy of *Aspergillus* infection is also significantly complicated by the increased risk of severe forms of aspergillosis in immunocompromised persons [7]. In this body of research, we will look at how *Aspergillus* infections affect patients with chronic conditions as well as how to avoid and treat them.

2. Clinical manifestations, diagnosis, and epidemiology of the spectrum of *Aspergillus* diseases

Lung infections brought on by *Ammophilus fumigatus* are produced by airborne conidia, which are present in both indoor and outdoor settings at concentrations ranging between 1 and 100 conidia per m^3, but which can reach up to 108 conidia per m^3 in some circumstances [8]. Because of this, *Aspergillus* spp. are regularly found in the respiratory tract cultures of asymptomatic patients who do not exhibit any symptoms of an invasive or allergic illness [9], and *Aspergillus* DNA has been found in 37% of lung biopsy samples from healthy persons. Additionally, *Aspergillus* colonization in up to 30% of people with chronic obstructive pulmonary disease (COPD) has been confirmed by culture [10]. Inadequate expression of the transcriptional factor ZNF77 in bronchial epithelia leads to defective epithelial cell integrity and upregulation of extracellular matrix (ECM) proteins that support conidial adhesion, which is the genetic basis of *Aspergillus* colonization, according to recent research [8]. Although *Aspergillus* colonization does not always result in infection, it increases the risk of invasive infection in several immunocompromised individuals. However, because invasive pulmonary aspergillosis arises from breathing *Aspergillus conidia*, environmental exposures affect the epidemiology of the disease. There have been several nosocomial clusters of invasive pulmonary aspergillosis (IPA) reported over the past

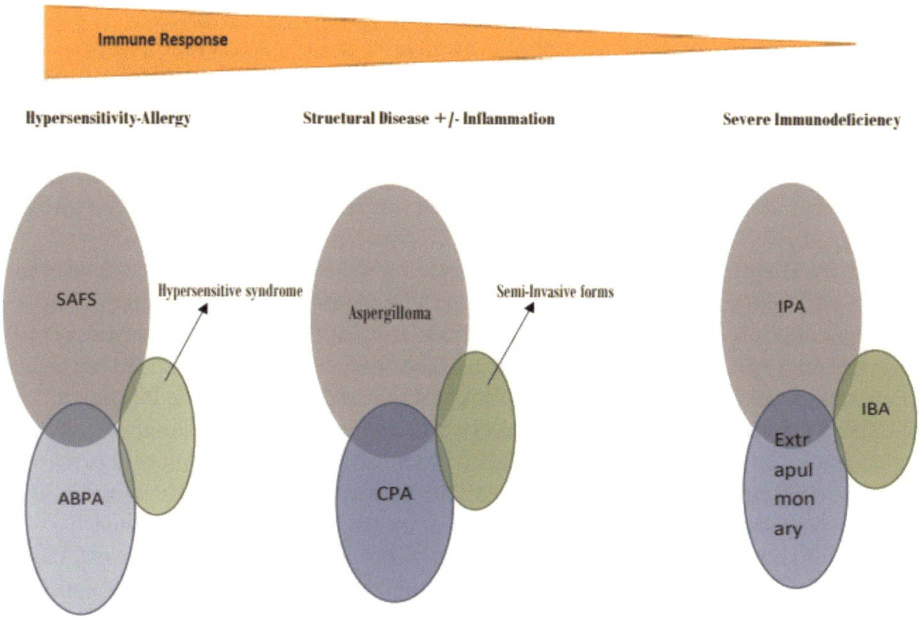

Figure 1.
Syndromes related to aspergillosis patients with various immune statuses.

three decades, which are typically associated with issues with hospital architecture and air-handling equipment [11].

Aspergillus species are responsible for a wide range of human ailments. Based on the underlying immunological status of the host, *Aspergillus* illnesses can be roughly categorized into three classes [8]. These three categories have various pathogenetic routes, clinical manifestations, and overlapping traits. These three groups are depicted in **Figure 1** based on their respective clinical importance. The sections that follow will discuss a few chronic illnesses and their connections to the *Aspergillus* species.

Among the syndromes related to aspergillosis patients with various immune statuses are allergy-related bronchial pulmonary aspergillosis (ABPA), chronic pulmonary aspergillosis (CPA), invasive pulmonary aspergillosis (IPA), and invasive bronchial aspergillosis (IBA) as shown in **Figure 1** [8].

3. Effect of *Aspergillus* infections on people with lung chronic disease

3.1 Asthma

The increased incidence and prevalence of asthma in the industrialized world have generated significant concerns and coordinated research efforts. Numerous theories to explain this significant change in public health have emerged as a result of the lack of clarity surrounding the etiopathogenesis of asthma. One of the most well-liked theories, the hygiene hypothesis, makes use of the fascinating interaction between the innate and acquired immune systems. According to this theory, the developed world's increased hygiene is a direct cause of the growth in asthma [12]. Even while this advancement has all but eliminated many infectious diseases like cholera, the lack of pathogen exposure in early children seems to cause allergies and asthma later in life.

The hygiene hypothesis states that any bacterium that triggers a sizable type 1 cytokine response may prevent the onset of asthma later in life even if no established list of illnesses that are "protective" has been created [13]. The emergence of Th1-type autoimmune illnesses and the anti-allergic benefits of robust Th2-cytokine-mediated parasite infections cannot exclusively be attributed to insufficient environmental pathogen exposure. The same cannot be stated for the growth of allergies and asthma. Many studies are currently devoted to figuring out how genetics affects allergies and asthma [14].

Inhalant allergens have a key role in triggering the airway inflammation seen in patients with allergic asthma. It is becoming increasingly clear that fungi are important inhalant allergens [13]. Numerous studies have linked asthma to the vast genus of spore-forming fungus known as *Aspergillus*. The "aspergillum" brush used to dispense holy water is referenced in the name of this fungus. All people inhale its spores, although a healthy, normal person is rarely affected by them [15]. However, the fungus spores get stuck in the thick, viscous secretions that are frequently present in the airways of asthmatics. To maintain this state, which commonly develops in atopic people and produces asthma, *Aspergillus* antigens are continuously breathed [3].

It is significant to mention that allergic broncho-pulmonary aspergillosis (ABPA) and chronic pulmonary aspergillosis (CPA) can confuse asthmatic patients' symptoms [16]. The primary cause of allergic bronchopulmonary aspergillosis, which is primarily linked to *Aspergillus fumigatus*, is hypersensitivity illness. The threatening type of aspergillosis is usually present in people with allergic bronchopulmonary aspergillosis [8]. Asthma sufferers are prone to respiratory infections brought on by the common mold genus *Aspergillus*. According to Seyedmousavi et al. [17], exposure to *Aspergillus* can cause allergic reactions in people with asthma, resulting in airway inflammation and constriction and making it harder to breathe.

The effects of *Aspergillus* infections on people with asthma can vary depending on the infection's intensity and the person's general health. *Aspergillus* infections may sporadically result in the development of allergic bronchopulmonary aspergillosis (ABPA), a chronic respiratory illness marked by symptoms like coughing, wheezing, and dyspnea, according to Chotirmall et al. [7]. When *Aspergillus* infections are severe, they can result in invasive aspergillosis, a condition that can be fatal and seriously harm the lungs. Kosmidis and Denning [18] claim that those with weakened immune systems, such as those undergoing chemotherapy or organ transplantation, are more likely to contract this condition.

Epithelial cells or alveolar macrophages cause inflammation as part of innate immunity's protective mechanisms. Toll-like receptors (TLRs), C-type lectin receptors (CLRs), and nucleic-binding oligomerization domain (NOD)-like receptors (NLRs) are examples of pattern recognition receptors that are used by innate immunity to identify fungi. *Aspergillus* stimulates these receptors, which activates cytokines and results in cellular and humoral immune responses. *Aspergillus* is protected and eliminated by T-helper cell type 1, while T-helper cell type 2 reacts by preventing its elimination. The severe inflammation seen in allergic bronchopulmonary aspergillosis, which results in eosinophilia, increased mucus formation, and IgE antibody production is caused by an excessive T-helper cell type 2 response. This inflammation leads to airway hypersensitivity, which aggravates asthma by causing bronchial obstruction [19, 20]. The clinical spectrum of *Aspergillus*-associated hypersensitivity respiratory diseases includes *Aspergillus*-induced asthma, allergic bronchopulmonary aspergillosis (ABPA), and allergy *Aspergillus* sinusitis (AAS). Hypersensitivity pneumonitis can also be brought on by *Aspergillus*, albeit this is often only seen in non-atopic persons.

Aspergillus-caused asthma has not yet received the attention it merits. Given the association between the mold *Aspergillus* and asthma, it is imperative to understand the frequency of *Aspergillus* sensitivity in asthmatic participants in each geographical area [21].

3.2 Arthritis

Gamaletsou et al. [22] found that *Aspergillus* infection can elevate inflammatory levels in the body, which can aggravate arthritic symptoms. Infections with *Aspergillus* can affect the respiratory system and result in symptoms, including coughing, wheezing, and shortness of breath. This may be particularly difficult for people with arthritis who already experience breathing issues due to joint pain and stiffness [23]. Furthermore, those with impaired immune systems who have arthritis are more likely to develop *Aspergillus* infections. For those who take immunosuppressive medicines to address their arthritic symptoms, this may be very challenging. Furthermore, *Aspergillus* infections can increase pain, stiffness, and mobility issues in those who already have arthritis by damaging their joints [22]. Due to the infection's potential to hinder the body's capacity to absorb medication, *Aspergillus* infections can also undermine the efficacy of arthritis treatments [24].

3.3 Cystic fibrosis

Cystic fibrosis (CF), the most common fatal genetically inherited disease that is caused by a mutation in a gene that encodes the CFTR protein, affects one in every 2400 live births in Caucasian cultures [25]. Two thousand different CFTR variations have so far been identified, with the most frequent being F508del, a single amino acid loss that accounts for about 70% of disorders [26]. Infants born today are expected to live into their fifth decade with CF, which has improved throughout a generation from a condition that often killed infants in their early years to one with a median lifespan of 28 years.

Ion fluxes and intracellular calcium homeostasis are compromised when the CFTR protein is absent from the cell membrane, where it acts as an ATP-driven chloride channel. Through a cycle of infection and exaggerated inflammation, thickened mucus forms in the airway epithelial cells, impeding the mucociliary clearance of inhaled pathogens and ultimately leading to respiratory failure [27, 28]. Continual lung infections and airway inflammation are the primary causes of mortality and morbidity in CF patients. To end this destructive cycle, the prevention and treatment of airway infection have been the cornerstones of clinical management; yet, respiratory failure brought on by chronic or recurrent infection still accounts for over 90% of fatalities [29]. The pathogenesis of respiratory decline has typically been studied through the role of bacterial pathogens like *Staphylococcus aureus*, *Pseudomonas aeruginosa*, *Haemophilus influenzae*, and the *Burkholderia cepacia* complex. The significance of *Aspergillus* species and other filamentous fungi in the pathogenesis of non-ABPA (allergic bronchopulmonary aspergillosis) respiratory illness in CF has received little attention, despite their frequent isolation in respiratory samples. However, it has become clearer that *Aspergillus fumigatus* may also be crucial for the CF lung [30]. People with CF are susceptible to respiratory infections brought on by the fungus *Aspergillus* [31].

Aspergillus infections can have a variety of impacts on CF patients, depending on the infection's severity and overall health. The following are a few of the implications of *Aspergillus* infections in people with CF: *Aspergillus* infections can cause a variety of respiratory symptoms, including coughing, wheezing, shortness of breath, chest pain, and fever. These symptoms can be extremely severe in CF patients because the underlying illness of CF already damages a person's lungs [18]. Again, *Aspergillus* infections can damage the lungs and impede their ability to function, leading to greater breathing issues and a worsening of CF symptoms. Lung damage caused by *Aspergillus* infections can occasionally be permanent [32]. Furthermore, *Aspergillus* infections can reduce the quality of life of CF patients by limiting their ability to perform routine tasks, attend school or work, and participate in social activities [33]. Infections with *Aspergillus* can also weaken immunity, raising the possibility of further respiratory infections in CF patients [34]. Additionally, treating *Aspergillus* infections in CF patients can be difficult and typically requires lengthy courses of antifungal medication. The infection may occasionally come back even after treatment [35].

3.4 Chronic obstructive pulmonary disease

People who have *Aspergillus* spp. growing in their airways and having chronic obstructive pulmonary disease (COPD) are often regarded as contaminants. Although it is unclear how common invasive pulmonary aspergillosis (IPA) is in this community, accumulating evidence suggests that individuals with severe COPD have an increased risk of contracting the condition [10]. According to certain estimates, COPD is the underlying illness in 1% of individuals with IPA. Given that tissue samples are infrequently taken before death in COPD patients, it may be difficult to make a definitive diagnosis of IPA. To make a diagnosis, a combination of clinical traits, radiographic findings—often from thoracic computed tomography scans—microbiological results, and occasionally serological data is used [36].

Chronic obstructive pulmonary disease (COPD), a progressive lung disorder that makes breathing challenging, can be brought on by several factors, including extended exposure to irritants like cigarette smoke [37]. A reaction to *Aspergillus* called allergic bronchopulmonary aspergillosis (ABPA) can occasionally strike patients with COPD. Wheezing, coughing, and shortness of breath are some of the symptoms of ABPA, which can worsen if left untreated [38].

4. Effects of *Aspergillus* infections on people living with other chronic infections

4.1 Hepatitis B and C viruses

Co-infection with *Aspergillus* may increase the pathogenicity of the hepatitis B and C viruses, leading to increased hepatocellular damage in those who are infected. There has been a hepatitis outbreak in some parts of Western India, which is characterized by jaundice, rapidly developing ascites, portal hypertension, and a high fatality rate. The illness was associated with eating maize that had been heavily contaminated with *Aspergillus flavus*. According to an analysis of contaminated samples, victims may have consumed 2–6 mg of aflatoxin per day over a month. Massive cells and bile duct development were found in a liver sample obtained during a necropsy. The sickness appears to be brought on by toxicosis [39]. Infections with *Aspergillus* can worsen preexisting liver damage and cause hepatic impairment, especially in those who also have hepatitis B or C viruses [40]. The detoxification and elimination of xenobiotics, such as the mycotoxins *Aspergillus* generates, depends on the liver [41]. Fungal infections like aspergillosis might worsen the immunosuppressive effects of the hepatitis B and C viruses. *Aspergillus* infection considerably raises the risk of severe or widespread infection in people with hepatitis B or C viruses, thereby exacerbating the immune impairment already present in these individuals. As a result, an *Aspergillus* co-infection may result in more severe symptoms and a longer time to recover [42]. A prominent therapeutic strategy for treating patients with hepatitis B and C viral infections is the use of antiviral medications. Antifungal medications used to treat *Aspergillus* infection may interact unfavorably or induce drug interactions in this case with some antiviral medications [43]. Furthermore, those with hepatitis B or C virus infections may have a higher chance of dying if they also have an *Aspergillus* infection. Due to a complex interplay of impaired liver function, immunosuppression, potential drug interactions, and increased disease severity, people who have concurrent hepatitis B or C virus and *Aspergillus* co-infection may have higher mortality rates than those who only have one disease [8].

4.2 Cancers

Infections account for the majority of deaths in patients with acute leukemia and lymphoma [44]. Host defenses have been breached [45]. The most prevalent condition in these patient populations is candidiasis, which is followed by *Aspergillus* spp.-related fungal infections [46]. For patients using immunosuppressive medications for illnesses including collagen vascular disorders, kidney transplants, and cardiac transplants, *Aspergillus* infections, which are the primary factors in their mortality when they exist, are extremely deadly [47].

Aspergillus infections in cancer patients can have a variety of effects depending on the infection's intensity and the patient's overall health. One of the most common types of *Aspergillus* infections in people with cancer is invasive aspergillosis, which occurs when the fungus enters the bloodstream and spreads to other parts of the body [48]. Other signs of this kind of virus include coughing, fever, chest pain, and shortness of breath. In extreme cases, invasive aspergillosis can result in organ failure and death [5]. Infections with *Aspergillus* can also cause additional conditions in cancer patients, including sinusitis, pneumonia, and skin infections. These infections can be difficult to recognize and treat, especially in people with weakened immune systems [49]. Chemotherapy or other cancer treatments raise the incidence of *Aspergillus* infections in cancer patients. To prevent these disorders, healthcare providers may advise antifungal medications or take other precautions, such as limiting mold exposure and preserving good hygiene practices [48].

4.3 HIV/AIDS

A study of 35,252 HIV patients in a national database estimated the incidence of invasive aspergillosis in AIDS to be 3.5 cases per 1000 person-years [50]. Aspergillosis was discovered in 0.43% of HIV patients, according to the findings of another database study of 38 million hospital diagnoses [50]. People with HIV/AIDS may be more susceptible to infections, such as aspergillosis, as a result of their suppressed immune systems [50]. Symptoms after exposure to *Aspergillus* might vary depending on how serious the illness becomes. Patients with weakened immune systems, such as those with HIV/AIDS, may experience more severe infection symptoms, including fever, coughing, shortness of breath, chest discomfort, and other respiratory symptoms.

According to Wang et al. [51], it may result in high mortality or serious lung injury. Neutropenia, which can arise as a result of HIV treatment, is a recognized risk factor for invasive aspergillosis [50]. Neutropenia or corticosteroid use is associated with nearly half of aspergillosis infections in HIV patients [51]. Additional risk factors for aspergillosis infection in HIV patients include concurrent Pneumocystis jirovecii (PCP) infection and a CD4 count of less than 50–100 cells/mm^3 [52].

Skin infections and infections of other organs, notably the brain, can arise from exposure to *Aspergillus* in addition to lung infections [51]. It wasn't until 2017 that researchers learned that HIV-positive patients might also get CPA and that their infection patterns were similar to those of HIV-negative people. Usually, CPA makes other respiratory conditions worse. Many people who have unexplained pulmonary tuberculosis (PTB), sometimes referred to as a smear or GeneXpert negative TB, actually have CPA instead of PTB [53], yet they are mistreated, and some of them will die as a result. According to Adams et al. [54], there is a considerable clinical and radiological overlap between CPA and subacute invasive aspergillosis.

Compared to other invasive fungal infections (IFIs), less is known, especially about the incidence and prognosis of aspergillosis in patients with HIV/AIDS [50]. This is thought to be the result of the difficulty in making a clinical diagnosis, which causes many aspergillosis patients to go untreated for the bulk of their lives. After death, aspergillosis is usually discovered [55]. The overall impact of this fungus on patients with HIV/AIDS is unknown as a result [50]. Patients with HIV/AIDS should make an effort to lower their risk of acquiring *Aspergillus* and other fungus-related illnesses. This requires practicing excellent cleanliness, avoiding areas with a lot of molds, and seeking immediate medical attention if any signs of infection emerge [56].

4.4 Diabetes

Diabetes patients experience more severe illnesses, a higher risk of infection, and a higher death rate when compared to the general population [57]. *Aspergillus fumigatus* (*A. fumigatus*) is the most common opportunistic airborne fungal infection that results in fatal invasive pulmonary aspergillosis (IPA) in immunocompromised individuals [8]. According to research by Ghanaat and Tayek [58], non-immunocompromised diabetic patients are more likely to acquire invasive aspergillosis. What makes diabetics more susceptible to an *A. fumigatus* infection is still a mystery, though. According to a study, people with diabetes had a poorer prognosis for fungal pneumonia since diabetes is an independent risk factor for long-term hospitalization for the condition [59]. Research on the effects of diabetes on pulmonary *A. fumigatus* infection uses a streptozotocin-induced mice model of diabetes as an example. A study revealed a more severe course of the pulmonary *A. fumigatus* infection in diabetic mice as demonstrated by a considerably poorer survival rate and clearance of *A. fumigatus* [60], in addition to the observed increased fungal burden 24 hours post-pulmonary *A. fumigatus* infection. Both the inflammatory and immunological responses are necessary for the host to be protected from pulmonary *A. fumigatus* infection. A good inflammatory response is necessary for the fungal infection to be eradicated. The overactive immune response that produces cytokines abruptly and in enormous numbers is known as hypercytokinemia, commonly referred to as a cytokine storm, and it can be even more destructive than the diseases that are invading the body [61]. When diabetic mice were infected with *A. fumigatus* in the lungs, the inflammatory response was hyperactive, as seen by notably extended and increased lung leukocyte infiltration as well as noticeably greater plasma cytokine expression.

The abnormal reaction was strongest in the early phases of infection. The inflammatory and immune responses, including cytokine-cytokine receptor interaction, tumor necrosis factor (TNF) signaling pathway, nucleotide-binding oligomerization domain-like (NOD-like) receptor, and toll-like receptor (TLR) signaling pathways, were the biological processes most enriched in diabetes. On the second day after infection, a transcriptome analysis of the lung tissue indicated this. Together, these results show that pulmonary *A. fumigatus* infection in diabetes results in a rapid, exaggerated inflammatory response that raises mortality. For diabetics, aspergillosis can be very problematic. Immunity can be weakened by diabetes, making it more challenging for the body to fight infections. Some diabetes medications may potentially increase the susceptibility of the immune system. Because of this, individuals with diabetes who are exposed to *Aspergillus* may get aspergillosis more frequently and with a more serious illness [60]. To lower their risk of aspergillosis and other fungal illnesses, diabetics must take precautions. This may require controlling blood sugar levels, keeping an impeccable standard of cleanliness, avoiding contact with mold and other fungi, and seeking medical attention as soon as any signs of aspergillosis or other illnesses appear. People with diabetes should work closely with their healthcare professionals to manage their condition and watch out for any complications, such as infections [62].

4.5 Obesity

Aspergillus infections are caused by a fungus of that name. These diseases can affect several parts of the body, including the skin, sinuses, and lungs. There is not any evidence yet that *Aspergillus* infections have a direct impact on weight. However, several studies suggest that several environmental factors, such as exposure to toxins and diseases, may have an impact on the development of obesity. For instance, exposure to endocrine-disrupting chemicals (EDCs) and other pollutants has been linked to increased adiposity in both human and animal models. *Aspergillus* infections may indirectly cause obesity by impairing the immune system and increasing susceptibility to other infections or variables in the environment that promote obesity [63]. It is critical to realize that obesity is a complex condition with a variety of underlying causes, including genetics, lifestyle, and environmental factors. Although *Aspergillus* infections may not directly contribute to obesity, they can nevertheless have major health consequences and must be treated effectively to prevent issues.

4.6 Alzheimer's disease

Alzheimer's disease is a neurological ailment that typically affects the elderly and gets worse with time. The hallmarks of this disease in the brain are amyloid plaques and neurofibrillary tangles, which result in neuronal cell death, vascular dysfunction, and inflammatory processes. In a study, researchers looked at whether people with Alzheimer's disease had fungal infections. Proteomic research provides substantial evidence that brain samples from Alzheimer's disease patients contain fungus-related proteins. Additionally, PCR analysis of these samples revealed a variety of fungal species, depending on the patient and the tissue investigated. Brain tissues included a variety of fungi, according to DNA research. Together, these results show that the brains of people with Alzheimer's disease contain fungus macromolecules. To our knowledge, these findings are the first evidence that fungi can be discovered in the brain tissues of Alzheimer's disease patients [64]. The specific impact of *Aspergillus* on people with Alzheimer's disease is unknown, though. Alzheimer's disease can make people more vulnerable to infections due to a weakened immune system, although there is no conclusive evidence linking it to an increased risk of *Aspergillus* infections [7]. However, patients with Alzheimer's disease may be more prone to developing respiratory infections, including those caused by *Aspergillus*, due to their weakened immune systems and difficulties swallowing, which can result in aspirating food or liquids into the lungs. Respiratory infections, according to Tangaleela et al. [65], can aggravate Alzheimer's disease symptoms and lead to consequences like pneumonia. Caretakers of individuals with Alzheimer's disease must take steps to prevent respiratory infections, including ensuring a clean environment, watching out for infection warning signs, and seeking medical attention as soon as symptoms develop. If an *Aspergillus* infection is suspected, a healthcare practitioner can perform diagnostic testing and recommend appropriate treatment options, such as antifungal medication [66].

4.7 Depression

Although some evidence points to *Aspergillus* infections as a possible cause of depression in some individuals, the relationship between the two conditions is convoluted and poorly understood. Continual *Aspergillus* infections have been linked to depressive symptoms such as low mood, loss of interest in activities, and low energy, according to research. The physical side effects of the infection, such as fatigue, soreness, and breathing issues, are likely what are aggravating these emotions [8]. It is likely that depressive disorders raise the risk of *Aspergillus* infections because they can weaken the immune system and make patients more susceptible to infections [67]. This is so that stress and other factors associated with depression can have this effect [67].

4.8 Stroke

Multiple organs could be affected by aspergillosis that has spread throughout the body. According to Santa-Ramrez et al. [68], the most serious side effect is brain infection, which occurs in 10–15% of patients and has a fatality rate of more than 90% even while receiving guided antifungal medication. A CNS infection may be acquired through hematogenous spread (typically from a pulmonary center), contiguous dissemination from a paranasal sinus infection, or direct iatrogenic injection during cerebral invasive procedures [69]. *Aspergillus* spp. CNS infection has been linked to numerous different clinical features. Up to 65% of these patients may have focal deficits Nevertheless, just one case report and not any clinical series have been used to describe the prevalence of clinical manifestations of acute stroke [69].

One of the infections that *Aspergillus*, a type of fungus, can cause is invasive aspergillosis, a serious condition that can affect people with weakened immune systems. It is unknown how aspergillosis impacts stroke, despite research suggesting that some illnesses, such as pneumonia and urinary tract infections, can increase the risk of the condition. According to case reports and small studies, aspergillosis may cause strokes, especially in people with compromised immune systems. The risk of stroke is increased by invasive aspergillosis, which can cause brain blood vessel inflammation and damage. Additionally, aspergillosis-related blood clots have been connected to a higher risk of stroke [70]. More research is required to completely understand the relationship between aspergillosis and stroke.

4.9 Osteoporosis

A disorder called osteoporosis is characterized by a loss of bone mass and a higher risk of fractures. There is little data on how *Aspergillus* infections affect osteoporosis, but some studies have found a link between osteoporosis and persistent *Aspergillus* lung infections. This is due to the possibility of persistent inflammation brought on by *Aspergillus* infections, which can promote bone resorption and reduce bone growth [71]. Additionally, corticosteroids and other antifungal drugs used to treat *Aspergillus* infections can raise the risk of osteoporosis. By inhibiting bone production and increasing bone resorption, corticosteroids are known to reduce bone density [72]. The link between *Aspergillus* infection and osteoporosis is complicated and can change depending on the person and the severity of the infection, and it is crucial to remember.

4.10 Chronic kidney disease

Aspergillus infections can be extremely difficult to cure and can have serious negative effects on the health of patients with chronic kidney disease (CKD). Immunosuppressive medications, which are frequently administered to decrease inflammation and lower the risk of rejection following kidney transplantation, are frequently linked to *Aspergillus* infections in CKD patients. These drugs may lower immunological function, increasing the risk of fungal infections in patients [73]. Patients with CKD may experience severe effects from *Aspergillus*, especially if the infection is not identified and treated right once. Invasive pulmonary aspergillosis (IPA), a potentially fatal illness that can cause lung damage, sepsis, and death, is brought on by *Aspergillus* [74]. In addition to the infection's direct effects, *Aspergillus* might aggravate the patient's underlying CKD and further jeopardize their health. Mycotoxins, which the fungus generates, can harm the kidneys and cause acute renal injury or worsen underlying CKD [41]. As many antifungal drugs can be harmful to the kidneys, treating *Aspergillus* infections in CKD patients can be difficult and necessitate close monitoring of the patient's renal function. To help the immune system more effectively combat the infection, immunosuppressive medicines may also need to be changed or stopped [75]. In conclusion, *Aspergillus* infections can have serious effects on the health of CKD patients, including deteriorating renal function and potentially fatal consequences. To lessen the effect of the infection on the patient's health, early identification and fast treatment are crucial [75].

4.11 Oral disease

A particular kind of fungus called *Aspergillus* can infect the human body, including the oral cavity. Various effects on oral health may result from these infections. Oral aspergillosis is one of the most prevalent *Aspergillus* infections of the oral cavity. This can happen to persons who have compromised immune systems, such as those who have cancer, HIV/AIDS, or other immune-compromising diseases. Additionally, it might happen to patients who have undergone particular dental treatments, like tooth extractions [76]. Several symptoms, such as discomfort, swelling, and redness in the afflicted area, can be brought on by oral aspergillosis. Along with making it difficult to speak or swallow, it can also lead to the emergence of sores or ulcers [77]. *Aspergillus* infections can contribute to the development or aggravation of other oral disorders in addition to causing direct harm to the oral cavity. For instance, *Aspergillus* infections can raise your risk of getting gum disease, periodontitis, and other oral infections [76].

4.12 Heart disease

Infections with *Aspergillus* primarily affect the respiratory system, but they can also spread to other bodily organs, such as the heart. For people with heart problems, *Aspergillus* infections can have severe negative effects. The fungus can inflame the heart muscle, which can result in cardiac failure, arrhythmias, and pain in the chest. Additionally, *Aspergillus* can infect the heart valves, which increases the risk of endocarditis and damages the heart valves [78]. *Aspergillus* infections are more likely to occur in those with compromised immune systems, such as those with HIV/AIDS or those receiving chemotherapy. But if they breathe in a lot of *Aspergillus* spores, even those with strong immune systems can have *Aspergillus* infections [8]. Antifungal medications are frequently used to treat *Aspergillus* infections in people with heart disease because they can assist to destroy the fungus and reduce inflammation. To correct heart valve damage brought on by the infection in some circumstances, surgery may be required [79]. By avoiding places where the fungus is prone to grow, such as damp or moldy locations, and by wearing protective masks when dealing with soil, compost, or other organic materials, people with heart disease should take precautions to prevent *Aspergillus* infections.

5. Conclusion

Aspergillus is a genus of fungi that can infect people with weakened immune systems, especially those who already have ongoing diseases. People with persistent infections may experience everything from minor symptoms to potentially fatal complications as a result of *Aspergillus* infections. Invasive pulmonary aspergillosis (IPA), which happens when the fungus penetrates the lungs and produces inflammation, is one of the most prevalent kinds of *Aspergillus* infections. This may result in symptoms including fever, coughing, pain in the chest, and shortness of breath. IPA can be particularly harmful in patients with chronic infections, since their compromised immune systems may not be able to successfully fight off the infection. Infections with *Aspergillus* can also affect the skin, nails, and sinuses in addition to other regions of the body. These infections may be more difficult to treat in patients with persistent infections and may call for more drastic measures, such as surgery or antifungal drugs. People with chronic infections may further endure psychological and emotional repercussions in addition to the physical signs and symptoms of *Aspergillus* infections. The ongoing risk of infection can be stressful and traumatic, and the necessity for repeated medical treatments can interfere with normal life. People with persistent infections may experience significant and wide-ranging impacts from *Aspergillus* infections. People with chronic infections need to take precautions to reduce their chance of contracting *Aspergillus* infections, such as maintaining excellent hygiene and avoiding mold and other environmental triggers. Additionally, it is crucial for medical professionals to keep an eye out for any indications of *Aspergillus* infections in this population and to offer fast and effective treatment when necessary. In conclusion, *Aspergillus* infections can affect patients with chronic illnesses in a variety of ways, ranging from medical symptoms to psychological distress. Healthcare professionals and patients alike should be aware of these hazards and take precautions to reduce them. Despite the difficulties presented by *Aspergillus* infections, persons with chronic infections can lead healthy, fulfilling lives with the right treatment and management, and also more research is need in the area of *Aspergillus* and aspergillosis in people living with chronic diseases to establish more mechanisms and associations between these fungi and other infections.

Author details

Bismark Dabuo, Nunekpeku Xorlali, Ndego Timothy Amoliga, Zyaara Kono Atibodu, Precious Mavis Newman, Alhassan Mohammed, Raymond Adongsakiya Ali and Abubakari Abudu*
University for Development Studies, Ghana

*Address all correspondence to: abuduabubakari688@gmail.com

References

[1] Merad Y, Derrar H, Belmokhtar Z, Belkacemi M. *Aspergillus* genus and its various human superficial and cutaneous features. Pathogens. 2021;**10**(6):1-17. DOI: 10.3390/pathogens10060643

[2] Haines SR, Siegel JA, Dannemiller KC. Modeling microbial growth in carpet dust exposed to diurnal variations in relative humidity using the "time-of-wetness" framework. Wiley. 2020;**30**(5): 978-992. DOI: 10.1111/ina.12686

[3] Samonis G, Koutserimpas C, Vrioni G, Martinez EK, Kouloumentas P, Alpantaki K, et al. Fungal septic knee arthritis caused by *aspergillus fumigatus* following anterior cruciate ligament reconstruction. Diagnostics. 2021;**11**(11): 1-7. DOI: 10.3390/diagnostics11111975

[4] Hammond EE, McDonald CS, Vestbo J, Denning DW. The global impact of *aspergillus* infection on COPD. BMC Pulmonary Medicine. 2020;**20**(1):1-10. DOI: 10.1186/s12890-020-01259-8

[5] Firacative C. Invasive fungal disease in humans: Are we aware of the real impact? Memorias Do Instituto Oswaldo Cruz. 2020;**115**(9):1-9. DOI: 10.1590/ 0074-02760200430

[6] Al Shakirchi M, Sorjonen K, Klingspor L, Bergman P, Hjelte L, de Monestrol I. The effects of *aspergillus fumigatus* colonization on lung function in patients with cystic fibrosis. Journal of Fungi. 2021;7(11):2-12. DOI: 10.3390/ jof7110944

[7] Chotirmall SH, Al-Alawi M, Mirkovic B, Lavelle G, Logan PM, Greene CM, et al. *Aspergillus*-associated airway disease, inflammation, and the innate immune response. BioMed Research International. 2013;**2013**(1):14. DOI: 10.1155/2013/723129

[8] Jean-Paul Latgé A, Chamilosa G. crossm. American Society of Microbiology. 2020;**33**(1):1-75

[9] Otu A, Kosmidis C, Mathioudakis AG, Ibe C, Denning DW. The clinical spectrum of aspergillosis in chronic obstructive pulmonary disease. Infection. 2023;**0123456789**:1-5. DOI: 10.1007/s15010-022-01960-2

[10] Shahi M, Ayatollahi Mousavi SA, Nabili M, Aliyali M, Khodavaisy S, Badali H. Aspergillus colonization in patients with chronic obstructive pulmonary disease. Current Medical Mycology. 2015;**1**(3):45-51. DOI: 10.18869/acadpub. cmm.1.3.45

[11] Gago S, Denning DW, Bowyer P. Pathophysiological aspects of aspergillus colonization in disease. Medical Mycology. 2019;**57**:S219-S227. DOI: 10.1093/mmy/myy076

[12] Ndlovu V, Chimbari MJ, Sibanda E. Assessing the nature of asthma in African epidemiological studies: A scoping review protocol. Systematic Reviews. 2020;**9**(1):1-7. DOI: 10.1186/ s13643-020-01491-7

[13] Milgrom H, Huang H. Allergic disorders at a venerable age: A mini-review. Gerontology. 2014;**60**(2):99-107. S. Karger AG. DOI: 10.1159/000355307

[14] Liu AH. Revisiting the hygiene hypothesis for allergy and asthma. Journal of Allergy and Clinical Immunology. 2015;**136**(4):860-865. DOI: 10.1016/j.jaci.2015.08.012

[15] Indranil S. Micology veterinary. In: Veterinary Mycology. Vol. 9, Issue 2. 2015. Available from: http://link.springer. com/10.1007/978-81-322-2280-4_3

[16] Denning DW, Pleuvry A, Cole DC. Global burden of allergic bronchopulmonary aspergillosis with asthma and its complication chronic pulmonary aspergillosis in adults. Medical Mycology. 2013;**51**(4):361-370. Oxford University Press. DOI: 10.3109/13693786.2012.738312

[17] Seyedmousavi S, Guillot J, Arné P, De Hoog GS, Mouton JW, Melchers WJG, et al. *Aspergillus* and aspergilloses in wild and domestic animals: A global health concern with parallels to human disease. Medical Mycology. 2015;**53**(8): 765-797. DOI: 10.1093/mmy/myv067

[18] Kosmidis C, Denning DW. The clinical spectrum of pulmonary aspergillosis. Thorax. 2015;**70**(3):270-277. DOI: 10.1136/thoraxjnl-2014-206291

[19] Becker KL, Gresnigt MS, Smeekens SP, Jacobs CW, Magis-Escurra C, Jaeger M, et al. Pattern recognition pathways leading to a Th2 cytokine bias in allergic bronchopulmonary aspergillosis patients. Clinical and Experimental Allergy. 2015;**45**(2):423-437. DOI: 10.1111/cea.12354

[20] Porsbjerg C, Menzies-Gow A. Co-morbidities in severe asthma: Clinical impact and management. Respirology. 2017;**22**(4):651-661. Blackwell Publishing. DOI: 10.1111/resp.13026

[21] Lal A, Akhtar J, Pinto S, Grewal H, Martin K. Recurrent pulmonary embolism and hypersensitivity pneumonitis secondary to *aspergillus*, in a compost plant worker: Case report and review of literature. Lung. 2018;**196**(5): 553-560. Springer New York LLC. DOI: 10.1007/s00408-018-0142-6

[22] Gamaletsou MN, Rammaert B, Bueno MA, Sipsas NV, Moriyama B, Kontoyiannis DP, et al. *Aspergillus* arthritis: Analysis of clinical manifestations, diagnosis, and treatment of 31 reported cases. Medical Mycology. 2017;**55**(3):246-254. DOI: 10.1093/mmy/myw077

[23] World Health Organization 2016. International Statistical Classification of Diseases and Related Health Problems, 10th Revision ICD-10 : Tabular List. World Health Organization, 1, 2, 33–95. Available from: http://www.who.int/classifications/icd/icdonlineversions/en/

[24] Du W, Gao Y, Liu L, Sai S, Ding C. Striking back against fungal infections: The utilization of nanosystems for antifungal strategies. International Journal of Molecular Sciences. 2021;**22**(18):2-35. DOI: 10.3390/ijms221810104

[25] Crull MR, Ramos KJ, Caldwell E, Mayer-hamblett N, Aitken ML, Goss CH. Change in Pseudomonas aeruginosa prevalence in cystic fibrosis adults over time. BMC Pulmonary Medicine. 2016; **16**(176):1-7. DOI: 10.1186/s12890-016-0333-y

[26] Lopes-Pacheco M. CFTR modulators: The changing face of cystic fibrosis in the era of precision medicine. Frontiers in Pharmacology. 2020;**10**(February):1-29. DOI: 10.3389/fphar.2019.01662

[27] Hanssens LS, Duchateau J, Casimir GJ. Cftr protein: Not just a chloride channel? Cell. 2021;**10**(11):1-15. DOI: 10.3390/cells10112844

[28] Robledo-Avila FH, Ruiz-Rosado J d D, Brockman KL, Kopp BT, Amer AO, McCoy K, et al. Dysregulated calcium homeostasis in cystic fibrosis neutrophils leads to deficient antimicrobial responses. The Journal of Immunology. 2018;**201**(7):2016-2027. DOI: 10.4049/jimmunol.1800076

[29] Ghigo A, Prono G, Riccardi E, De Rose V. Dysfunctional inflammation in

cystic fibrosis airways: From mechanisms to novel therapeutic approaches. International Journal of Molecular Sciences. 2021;**22**(4):1-23. DOI: 10.3390/ijms22041952

[30] McDermott G, Reece E, Renwick J, Dublin TC. Microbiology of the cystic fibrosis airway. Encyclopedia of Microbiology. 2019;**2002**:1-14. DOI: 10.1016/B978-0-12-809633-8.90263-2

[31] Burgel PR, Paugam A, Hubert D, Martin C. *Aspergillus fumigatus* in the cystic fibrosis lung: Pros and cons of azole therapy. Infection and Drug Resistance. 2016;**9**:229-238. DOI: 10.2147/IDR.S63621

[32] Lv Q, Elders BBLJ, Warris A, Caudri D, Ciet P, Tiddens HAWM. *Aspergillus*-related lung disease in people with cystic fibrosis: Can imaging help us to diagnose disease? European Respiratory Review. 2021;**30**(162):1-14. DOI: 10.1183/16000617.0103-2021

[33] Magee LC, Louis M, Khan V, Micalo L, Chaudary N. Managing fungal infections in cystic fibrosis patients: Challenges in clinical practice. Infection and Drug Resistance. 2021;**14**:1141-1153. DOI: 10.2147/IDR.S267219

[34] Curran AK, Hava DL. Allergic diseases caused by *aspergillus* species in patients with cystic fibrosis. Antibiotics. 2021;**10**(4):1-13. DOI: 10.3390/antibiotics10040357

[35] Poore TS, Hong G, Zemanick ET. Fungal infection and inflammation in cystic fibrosis. Pathogens. 2021; **10**(5):1-13. DOI: 10.3390/pathogens10050618

[36] Fréalle E, Reboux G, Le O, Bautin N. Impact of domestic mould exposure on *aspergillus* biomarkers and lung function in patients with chronic obstructive pulmonary disease. Environmental Research. 2021;**195**:1-25. DOI: 10.1016/j.envres.2021.110850

[37] Singh J, Kumar M, Sharma A, Pandey G, Chae K, Lee S. 2016. We are IntechOpen, the world's leading publisher of Open Access books Built by scientists , for scientists TOP 1%. Intech, 11(tourism), 13. Available from: https://www.intechopen.com/books/advanced-biometric-technologies/liveness-detection-in-biometrics

[38] Sockrider M, Wenzel SE, Castro M, Kulkarni H. What is allergic bronchopulmonary aspergillosis (ABPA)? American Journal of Respiratory and Critical Care Medicine. 2014;**190**(6):P3-P4. DOI: 10.1164/rccm.1906P3

[39] Benkerroum N. Aflatoxins: Producing-molds, structure, health issues and incidence in southeast asian and sub-saharan african countries. International Journal of Environmental Research and Public Health. 2020; **17**(4):2-40. DOI: 10.3390/ijerph17041215

[40] Zhai XR, Tong JJ, Wang HM, Xu X, Mu XY, Chen J, et al. Infection deteriorating hepatitis B virus related acute-on-chronic liver failure: A retrospective cohort study. BMC Gastroenterology. 2020;**20**(1):1-13. DOI: 10.1186/s12876-020-01473-y

[41] Navale V, Vamkudoth KR, Ajmera S, Dhuri V. *Aspergillus* derived mycotoxins in food and the environment: Prevalence, detection, and toxicity. Toxicology Reports. 2021;**8**:1008-1030. DOI: 10.1016/j.toxrep.2021.04.013

[42] Roudbary M, Kumar S, Kumar A, Černáková L, Nikoomanesh F, Rodrigues CF. Overview on the prevalence of fungal infections, immune response, and

microbiome role in COVID-19 patients. Journal of Fungi. 2021;**7**(9):1-28. DOI: 10.3390/jof7090720

[43] Frediansyah A, Sofyantoro F, Alhumaid S, Mutair A Al, Albayat H, Altaweil HI, et al. Microbial natural products with antiviral activities, including anti-SARS-CoV-2: A review. Molecules. 2022;**27**(13):1-27. DOI: 10.3390/molecules27134305

[44] Logan C, Koura D, Taplitz R. Updates in infection risk and management in acute leukemia. Hematology (United States). 2020;**20**(1):135-139. DOI: 10.1182/HEMATOLOGY.2020000098

[45] Tan DX, Hardeland R. Targeting host defense system and rescuing compromised mitochondria to increase tolerance against pathogens by melatonin may impact outcome of deadly virus infection pertinent to COVID-19. Molecules. 2020;**25**(19):1-26. DOI: 10.3390/molecules25194410

[46] Vazquez JA, Miceli MH, Alangaden G. Invasive fungal infections in transplant recipients. Therapeutic Advances in Infectious Disease. 2013;**1**(3):85-105. DOI: 10.1177/2049936113491936

[47] Davies S, Guidry C, Politano A, Rosenberger L, Mcleod M, Hranjec T, et al. *Aspergillus* infections in transplant and non-transplant surgical patients. Surgical Infections. 2014;**15**(3):207-212. DOI: 10.1089/sur.2012.239

[48] Mousset S, Buchheidt D, Heinz W, Ruhnke M, Cornely OA, Egerer G, et al. Treatment of invasive fungal infections in cancer patients - Updated recommendations of the infectious diseases working party (AGIHO) of the German Society of Hematology and Oncology (DGHO). Annals of Hematology. 2014;**93**(1):13-32. DOI: 10.1007/s00277-013-1867-1

[49] Chamilos G. crossm. American Society of Microbiology. 2020;**33**(1):1-75

[50] Denning DW, Morgan EF. Quantifying deaths from aspergillosis in HIV positive people. Journal of Fungi. 2022;**8**(11):2-14. DOI: 10.3390/jof8111131

[51] Wang RJ, Miller RF, Huang L. Approach to fungal infections in HIV-infected individuals: Pneumocystis and beyond. Physiology & Behavior. 2017;**38**(3):465-477. DOI: 10.1016/j.ccm.2017.04.008.Approach

[52] Langlois ME, Lorillou M, Ferry T, Chidiac C, Valour F. Cystic lung lesions revealing a pneumocystis jirovecii and *aspergillus* flavus co-infection in an HIV-infected patient. International Journal of Infectious Diseases. 2015;**37**:e143–e144. Elsevier. DOI: 10.1016/j.ijid.2015.06.021

[53] Ocansey BK, Otoo B, Adjei A, Gbadamosi H, Kotey FCN, Kosmidis C, et al. Chronic pulmonary aspergillosis is common among patients with presumed tuberculosis relapse in Ghana. Medical Mycology. 2022;**60**(9):1-7. DOI: 10.1093/mmy/myac063

[54] Adams B, Nunes JM, Page MJ, Roberts T, Carr J, Nell TA, et al. Parkinson's disease: A systemic inflammatory disease accompanied by bacterial inflammagens. Frontiers in Aging. Neuroscience. 2019;**10**(JUL):1-17. DOI: 10.3389/fnagi.2019.00210

[55] Russo A, Tiseo G, Falcone M, Menichetti F. Pulmonary aspergillosis: An evolving challenge for diagnosis and treatment. Infectious Diseases and Therapy. 2020;**9**(3):511-524. DOI: 10.1007/s40121-020-00315-4

[56] Oliveira M, Oliveira D, Lisboa C, Boechat JL, Delgado L. Clinical manifestations of human exposure to Fungi. Journal of Fungi. 2023;**9**(3):1-41. DOI: 10.3390/jof9030381

[57] Erener S. Diabetes, infection risk and Covid-19. Molecular Metabolism. 19 Jun 2020:101044. DOI: 10.1016/j.molmet.2020.101044

[58] Ghanaat F, Tayek JA. HHS public access. HHS Public Access. 2018;**14**(5): 296-301. DOI: 10.4172/clinical-practice.1000125.Weight

[59] Berbudi A, Rahmadika N, Tjahjadi AI, Ruslami R. Type 2 diabetes and its impact on the immune system. Bentham Science. 2019;**16**(5):442-449. DOI: 10.2174/1573399815666191024085838

[60] Khanam A, Hithamani G, Naveen J, Pradeep SR, Barman S, Srinivasan K. Management of Invasive Infections in diabetes mellitus: A comprehensive review. Biologics. 2023;**3**(1):40-71. DOI: 10.3390/biologics3010004

[61] Palmieri B, Manenti A, Galotti F, Vadalà M. Innate immunity stimulation during the COVID-19 pandemic: Challenge by Parvulan. Hindawi. 2022; **2022**(4593598):9

[62] Lao M, Li C, Li J, Chen D, Ding M, Gong Y. Opportunistic invasive fungal disease in patients with type 2 diabetes mellitus from southern China: Clinical features and associated factors. Journal of Diabetes Investigation. 2020;**11**(3): 731-744. DOI: 10.1111/jdi.13183

[63] Gupta R, Kumar P, Fahmi N, Garg B, Dutta S. 2020. Since January 2020 Elsevier has created a COVID-19 resource centre with free information in English and Mandarin on the novel coronavirus COVID-19. The COVID-19 resource centre is hosted on Elsevier Connect , the company 's public news and information. Elsevier, January

[64] Pisa D, Alonso R, Rábano A, Rodal I, Carrasco L. Different brain regions are infected with Fungi in Alzheimer's disease. Scientific Reports. 2015;**5**:1-13. DOI: 10.1038/srep15015

[65] Thangaleela S, Sivamaruthi BS, Kesika P, Bharathi M, Chaiyasut C. Nasal microbiota, olfactory health, neurological disorders and aging—A review. Microorganisms. 2022;**10**(7):1-21. DOI: 10.3390/microorganisms10071405

[66] Australian Commission for Safety and Quality in Healthcare. Australian Guidelines for the Prevention and Control of Infection in Healthcare. National Health Medical Research Council; 2019. MAGICapp platform. Available from: http://www.legislationreview.nhmrc.gov.au/_files_nhmrc/publications/attachments/cd33_infection_control_healthcare.pdf

[67] Cañas-González B, Fernández-Nistal A, Ramírez JM, Martínez-Fernández V. Influence of stress and depression on the immune system in patients evaluated in an anti-aging unit. Frontiers in Psychology. 2020;**11**(August):1-10. DOI: 10.3389/fpsyg.2020.01844

[68] Santana-Ramírez A, Esparza-Gutiérrez S, Avila-Rodríguez P, Jiménez-Gómez J, Vélez-Gómez E, Bañuelos-Gallo D. Aspergillosis of the central nervous system in a previously healthy patient that simulated Creutzfeldt-Jakob disease. Surgical Neurology International. 2016;**7**(40):S940-S946. DOI: 10.4103/2152-7806.195230

[69] Winterholler M, Coras R, Geißdörfer W, Rammensee R, Gölitz P, Bogdan C, et al. Fatal mycotic aneurysm

of the basilar artery caused by *aspergillus fumigatus* in a patient with pituitary adenoma and meningitis. Frontiers in Medicine. 2017;**4**(JUL):1-7. DOI: 10.3389/fmed.2017.00113

[70] Kumar S, Muthu V, Bansal YS, Mehta N, Arora V. Invasive pulmonary aspergillosis with brain dissemination in an immunocompetent host. Autopsy and Case Reports. 2021;**11**:1-6. DOI: 10.4322/acr.2021.280

[71] Zou X, Ma Z, Liu X, Zhang K, Qiu C, Liang R, et al. Risk factors and prognostic value of osteoporosis in hospitalized patients with bronchiectasis. BMC Pulmonary Medicine. 2023;**23**(1):1-10. DOI: 10.1186/s12890-023-02346-2

[72] Panday K, Gona A, Humphrey MB. Medication-induced osteoporosis: Screening and treatment strategies. Therapeutic Advances in Musculoskeletal Disease. 2014;**6**(5):185-202. DOI: 10.1177/1759720X14546350

[73] Sato N, Yokoi H, Ichioka M, Ishii A, Matsubara T, Yanagita M. Invasive aspergillosis in the patient with focal segmental glomerulosclerosis initiating hemodialysis: A case report and mini-review. Renal Replacement Therapy. 2022;**8**(1):1-9. DOI: 10.1186/s41100-022-00455-y

[74] Levesque E, Ait-Ammar N, Dudau D, Clavieras N, Feray C, Foulet F, et al. Invasive pulmonary aspergillosis in cirrhotic patients: Analysis of a 10-year clinical experience. Annals of Intensive Care. 2019;**9**(1):2-9. DOI: 10.1186/s13613-019-0502-2

[75] Zhou L, Zhao H, Chen Z, Zhu L. Urinary tract aspergillosis in a patient with chronic kidney disease. BMJ Case Reports. 2017;**2017**:2-4. DOI: 10.1136/bcr-2017-221638

[76] Bharatha A, Kandamaran L, Krishnamurthy K. Commentary: "Fungal infections of Oral cavity: Diagnosis, management, and association with COVID-19.". SN Comprehensive Clinical Medicine. 2021;**3**(9):2002-2003. DOI: 10.1007/s42399-021-01017-9

[77] Poulopoulos A, Papadopoulos P, Andreadis D. Chemotherapy: Oral side effects and dental interventions. A review of the literature. Stomatological Disease and Science. 2017;**1**(2):35-49. DOI: 10.20517/2573-0002.2017.03

[78] Sulik-Tyszka B, Kacprzyk P, Mądry K, Ziarkiewicz-Wróblewska B, Jędrzejczak W, Wróblewska M. Aspergillosis of the heart and lung and review of published reports on fungal endocarditis. Mycopathologia. 2016;**181**(7–8):583-588. DOI: 10.1007/s11046-016-0012-9

[79] Patterson TF, Thompson GR, Denning DW, Fishman JA, Hadley S, Herbrecht R, et al. Practice guidelines for the diagnosis and management of aspergillosis: 2016 update by the infectious diseases society of America. Clinical Infectious Diseases. 2016;**63**(4):e1-e60. DOI: 10.1093/cid/ciw326

Section 2

Virulence and Pathogenesis

Chapter 4

Virulence Attributes in *Aspergillus fumigatus*

María Guadalupe Frías-De-León, Eduardo García-Salazar and Gustavo Acosta-Altamirano

Abstract

Aspergillus fumigatus is one of the most important opportunistic fungal pathogens. It causes various types of infections in humans, from skin, lung, and allergic infections to invasive infections. However, these stand out because their mortality rate can reach up to 95%. *A. fumigatus* is a ubiquitous fungus and, therefore, humans are in constant contact with it without major risk, except when there is a predisposing factor on the host, that allows the fungus to penetrate and invade the tissues. It is fascinating how this fungus manages to go from harmless to pathogenic as, in addition to the predisposing factors of the human, multiple attributes of the fungus intervene that favor its growth and survival in the host. Among these virulence attributes are thermotolerance, the ability to evade the immune response, some components of the cell wall, the production of secondary metabolites, compliance with nutritional requirements, and the production of melanin, among others. Furthermore, some of these virulence attributes are interrelated, making understanding the pathogenesis of aspergillosis more complex. This chapter presents a review of some virulence attributes that are known, to date, in *A. fumigatus*.

Keywords: *A. fumigatus*, virulence, pathogenesis, invasive aspergillosis, pathogenicity attributes

1. Introduction

The genus *Aspergillus* groups more than 200 species of filamentous fungi, with *A. fumigatus* being one of the most abundantly distributed species in the environment [1]. *Aspergillus* taxonomy is very complex, there are more and more species recognized as pathogenic for humans; however, *A. fumigatus* remains the most important species (Appendix 1). Being a ubiquitous and saprophytic fungus, it can grow and reproduce easily on decaying organic matter, soil, and dust in the air [2]. Until a few years ago, asexual reproduction was the only one recognized in *A. fumigatus*; it is now known to important species reproduce sexually also [3]. In the environment, this fungus reproduces mainly asexually, producing large numbers of small conidia (2–3 μm, the ideal size to go deep into the pulmonary alveoli) in structures called conidiophores, which disperse over great distances [2]. The conidia present in the air are constantly inhaled by humans (up to 5000 conidia per day), which can easily overcome the mucociliary

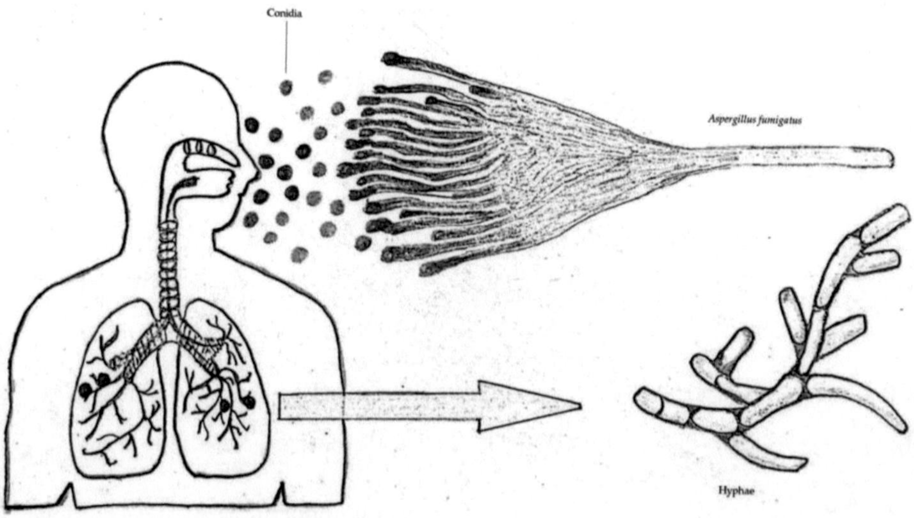

Figure 1.
Route of entry of A. fumigatus *to the host.*

clearance and reach the epithelial cells of the airways, where it begins to colonize and, after approximately 24 h of hyphal growth, produce some secondary metabolites that break the endothelial epithelium (**Figure 1**) [4, 5].

After damaging the epithelial layer of the alveoli, the fungus can enter the endothelium of blood vessels [6]. The pathogenesis of the disease mediated by *A. fumigatus* occurs in a multi-step manner and involves the morphological transition of the inhaled fungal spore to a hyphae form. Epithelial damage can be considered to occur in the early (conidia) or late (hyphae) phase of the fungal interaction with epithelial cells [7]. In healthy immunocompetent individuals, the growth of hyphae is impeded by immune mechanisms, whereas, in people with immune deterioration, such as those who have leukemia or who have undergone bone marrow or solid organ transplantation, mycelial development leads to severe disease, which can be life-threatening (**Figure 2**) [8–11].

The respiratory tract is the main route of entry and site of infection of *A. fumigatus*; however, in both immunocompetent and immunocompromised hosts, other sites such as the skin, peritoneum, kidneys, bones, eyes, and gastrointestinal tract can be infected [4]. Lung diseases caused by *A. fumigatus* are classified according to the site affected within the respiratory tract and the degree of mycelial colonization or invasion, both of which are influenced by the immune status of the host [4]. Thus, repeated exposure to *Aspergillus* conidia or antigens without mycelial colonization can lead to allergic diseases, such as asthma, allergic sinusitis, and alveolitis. Patients usually show improvement when the environmental source of exposure is removed [11]. On the other hand, when there is fungal colonization and mycelial growth in the host, allergic bronchopulmonary aspergillosis (ABPA), aspergilloma, and invasive aspergillosis (IA) can develop, and patients usually require therapeutic intervention to achieve clinical improvement. Unfortunately, this is not always achieved, particularly in cases of IA, where the mortality rate is high (80–90%) [10–12]. Therefore, the pathogenicity of *A. fumigatus* depends not only on the host's immune status but also on the ability of the fungus to adapt to the host environment, that is, on the virulence of the fungus strain [13]. Contrary to what happens in most primary pathogens, in which

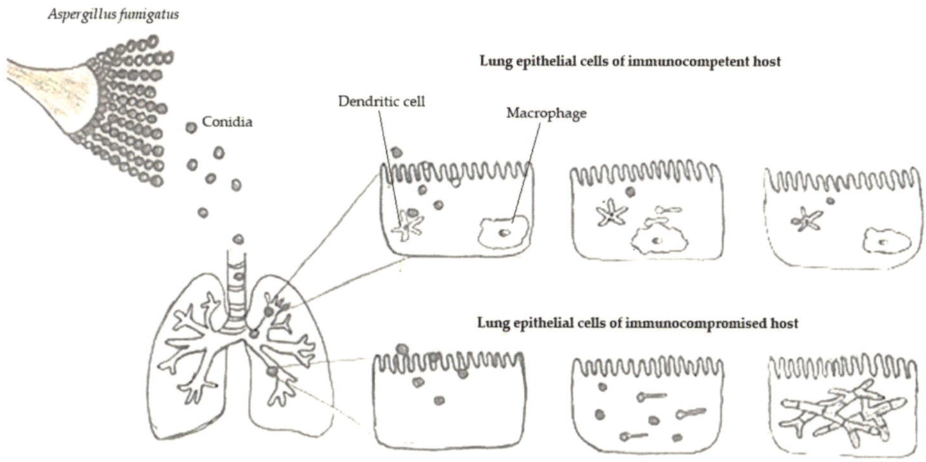

Figure 2.
A. fumigatus *host interaction and pathogenesis.*

virulence traits develop in association with the host [14], the virulence in *A. fumigatus* is multifactorial and determined by a series of attributes that are under polygenetic control [13, 15, 16]. The virulence attributes that contribute to the pathogenicity of *A. fumigatus* are related to various processes such as thermotolerance, cell surface organization, adhesion molecules present on the conidial surface, production of secondary metabolites, compliance with nutritional requirements, interactions with the host immune system and stress response [16, 17]. Although some of these attributes do not fit the classical definition of a virulence factor, they are essential in the pathogenesis of aspergillosis. Thus, their knowledge may provide new opportunities for developing antifungals [14]. Below are some of the main virulence attributes in *A. fumigatus*.

2. Thermotolerance

A. fumigatus is a thermotolerant fungus that can grow at high temperatures (55°C) and even survive up to 75°C [18]. Thermotolerance is an essential characteristic of the fungus. It allows it to carry out its primary functions, degrade organic matter, exceed the thermal exclusion barrier of mammals, including humans, and cause infections [19, 20]. Therefore, genes that regulate thermotolerance are considered virulence attributes. Some genes (*thtA, afpmt1, cgrA*) associated with thermotolerance have been described in *A. fumigatus* [21–23]. These genes have different functions; for example, the *afpmt1* and *thtA* genes contribute to fungal growth at 37 and 48°C, respectively [21, 22]. It has been seen that the deletion of these genes does not modify the virulence of the fungus. However, they indirectly influence pathogenicity, as they provide the ability to grow and persist within the human host, overcoming the thermal exclusion barrier. In the case of the *cgrA* gene, which is involved in the biogenesis of ribosomes at 37°C, it has been reported to have a direct influence on virulence since the mutant isolates produced by the deletion of this gene have a phenotype of low pathogenicity [23]. On the other hand, for *A. fumigatus* to survive human body temperature and transition from conidia to hyphae, it must resist high-temperature induced proteotoxic stress and flow in protein production demand [20]. Under these conditions, the so-called heat shock proteins (HSPs) become relevant. These proteins are upregulated with increasing temperature and stress induction, function as chaperones to facilitate proper folding and modification of proteins, and are molecules conserved between organisms. HSPs facilitate the acquisition of thermotolerance and allow human fungal pathogens to grow at human body temperature and survive after heat stroke at elevated temperatures. Several of these chaperones are necessary for morphological changes. Hsp90 and Hsp70 in human fungal pathogens contribute extensively to thermotolerance, morphological changes necessary for virulence, and tolerance to antifungal drugs [20]. In *A. fumigatus*, Hsp90 is involved in hyphae formation, so it has been suggested as a potential target of antifungal drugs [24]. Hsp70s are located on the cell surface of *A. fumigatus*, have high levels of expression at elevated temperatures, and influence morphological transitions [20, 25].

3. Adherence

After inhalation of the airborne conidia, adherence of *A. fumigatus* to host epithelial cells is essential for developing infection [16, 26]. After the conidia attach to the epithelial cells and to the extracellular matrix exposed in the airways, these cells recognize and internalize them. Some conidia manage to survive, avoiding the action of immune cells [27]. Recognition is achieved through the expression of different pattern recognition receptors (PRRs), including Toll-like receptors (TLRs), C-type lectin (CLR), and nucleotide-binding oligomerization domain (NOD)-like receptors (NLRs) that recognize fungal cell wall polysaccharides or fungal pathogen-associated molecular patterns (PAMPs) [28]. Recognizing specific PAMPs promotes the activation of antimicrobial mechanisms that help eliminate fungi. The most critical PAMPs of filamentous fungi are mannan, β-glucan, and chitin [28, 29]. During infection, the β-1,3 glucans of the cell wall are relevant PAMPs recognized by dectin-1, a type C lectin. Some studies have shown that hydrophobins hide these glucans in conidia but are exposed when the conidia swell and begin to germinate [30, 31]. After germination, exposure to β-1,3 glucan decreases due to hyphae-associated galactosaminogalactan (GAG) production, impairing the recognition of hyphae by dectin-1 [32]. There is evidence that GAG plays a vital role in mediating the adherence of hyphae to host epithelial cells. For example, GAG-deficient mutant isolates (Δuge3 and ΔmedA) reduce their adherence to lung epithelium cells of cell line A549 [27, 33]. On the other hand, the adhesion of the hyphae of the mutant Δuge3 improves notoriously when the purified GAG adheres directly to the epithelial cells A549 [32]. Thus, nullified GAG production in the mutant Δuge3 is associated with increased exposure to β-1,3 glucan in hyphae [34, 35]. This results in increased recruitment of leukocytes during lung infection and increased binding of dectin-1 to the surface of hyphae, leading to increased production of inflammatory cytokines by dendritic cells *in vitro* [31]. Inflammation and ciliary damage result in decreased mucociliary elimination, which in turn can favor adherence and germination of *A. fumigatus* conidia [26, 36].

4. Compliance with nutritionals requirements

4.1 Copper

Another virulence attribute recognized in *A. fumigatus* is copper [37]. The fungal hyphae must acquire copper from the environment and maintain the intracellular concentration within the micromolar range. In general, copper is internalized through high-affinity uptake systems, depending on the copper concentration in the extracellular medium [38, 39]. One of the high-affinity uptake systems comprises the family of copper transporter proteins (Ctr) associated with the membrane. They are small proteins (18–30 kDa) with up to three transmembrane domains and have the Cu + ion as a substrate [40–42]. Copper-binding motifs (Mets) located in the extracellular N-terminal region or transmembrane domain are rich in methionine (MxxM, MxM, or MxxxM) [43]. The MxxxM motif is essential in the transmembrane transport of Cu+ [40]. Ctr proteins are assembled to create a pore by which the transmembrane passage of Cu + is driven. Entry is facilitated when the intracellular concentration of the metal is low [44]. Copper in its Cu2+ oxidation state can also be internalized but must be reduced first to Cu + by the action of reductases present in the plasma membrane [39].

On the other hand, copper uptake is a strictly controlled process. When the intracellular level of Cu + exceeds the toxicity threshold, in addition to generating reactive oxygen species (ROS), the mechanism of detoxification or ion sequestration is activated to restore the balance of cellular copper. This mechanism is directed by the transcription factor AceA [37, 45]. The DNA-binding domain, known as "Cu_FIST," and the numerous cysteine residues arranged in CxC-CxxC segments throughout the protein sequence are the characteristic domains that identify this transcription factor. Within the DNA-binding domain, two different motifs are involved in binding stabilization. When there is an excess of copper, four Cu + atoms bind to the Cys-rich domain, causing a conformational change that facilitates AceA-DNA bonding. This binding allows transcription of the crpA coding gene for P-type ATPase. P-type-ATPase CrpA is synthesized in the endoplasmic reticulum and migrates to the plasma membrane, where CrpA pumps Cu + ions out of the cell to restore balance and reduce copper toxicity. AceA also activates the sod1 and cat1/2 genes, which encode superoxide dismutase (Sod1) and catalase (Cat1/2). Sod1 and Cat1/2 enzymes neutralize ROS generated by Cu + toxicity and even those generated by the host's defense mechanisms (**Figure 3**) [37, 45]. Therefore, the mechanism of copper detoxification is a critical factor in the viability of the pathogen during infection. In addition, both Sod1 and Cat1/2 also play an essential role in fungal virulence. Interestingly, host organisms have developed defense strategies against copper-dependent fungal pathogens. For example, by removing all Cu + in the infection area, copper deprivation in the pathogen can be induced, and its growth can be limited. Likewise, the copper mobilization mediated by the host's innate immune cells to the tissue invaded by the fungus is another defense mechanism that seeks to cause fungal poisoning by excess copper [46].

On the other hand, copper also acts as a cofactor of the laccases AfAbr1 and AfAbr2, which are involved in the melanin biosynthesis in *A. fumigatus* and,

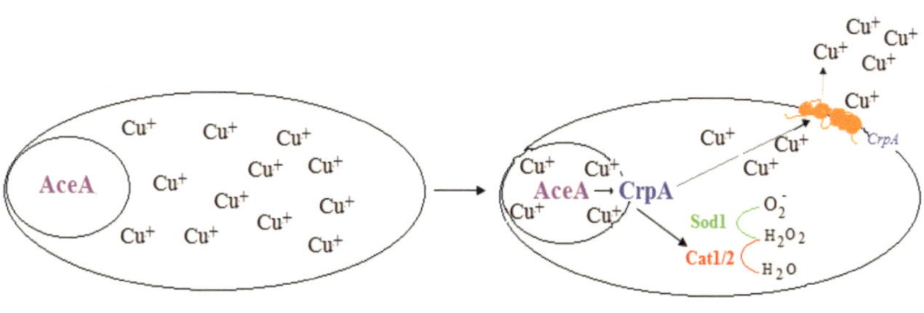

Figure 3.
Copper homeostasis in A. fumigatus.

therefore, in virulence [47]. Melanin confers a non-immunogenic state to the fungus. Therefore, if the absorption of copper in the hyphae decreases, the laccase activity is also reduced, generating melanin deficiency in the conidia, which makes them immunoreactive [48, 49]. The latter highlights the importance of copper uptake for pathogen viability within the host.

4.2 Iron

Iron is an essential nutrient for all living organisms, including *A. fumigatus* [50]. This fungus uses iron as a cofactor for fundamental biochemical activities, such as oxygen transport, energy metabolism, and DNA synthesis [51]. Therefore, during infection, iron competition is a crucial event that determines the outcome of the host-pathogen relationship [52]. While the host's innate immune system sequesters iron to reduce its free concentration to low levels and limit availability for *A. fumigatus*, the fungus uses different strategies to adapt to low environmental iron concentration [53], and even iron overload, as it is vital to maintain homeostasis. Two transcription factors maintain iron homeostasis: GATA factor SreA and factor bZIP HapX [51, 54]. When the fungus has enough iron for cellular activities, factor SreA represses iron absorption mediated by iron-reducing assimilation and siderophores to avoid toxic effects [55]. When iron is scarce, the factor HapX activates siderophore-mediated iron acquisition and, at the same time, saves iron by preventing its consumption in activities such as heme biosynthesis and respiration. Recent research has revealed that siderophores represent essential virulence factors contributing to a host's microbiome-metabolome dialog [56, 57]. HapX deficiency, but not SreA, attenuates the virulence of *A. fumigatus* in murine models of aspergillosis [55, 58], emphasizing the crucial role of iron-limiting adaptation in pathogenicity. It has been reported that immunocompromised patients with iron overload, after a transplant, have a high risk of developing invasive aspergillosis. Therefore, it is proposed that the pharmacological inhibition of the siderophores synthesis of *A. fumigatus* or chelating agents could favor this type of patient [52].

4.3 Zinc

In the same way as iron, zinc is an essential cofactor for many crucial metabolic processes, including the growth and virulence of *A. fumigatus* [59]. Therefore, the fungus has also developed mechanisms to capture zinc through transporters (zrfC) and maintains homeostasis through the action of a transcription factor (zafA). Thus, mutant strains lacking genes encoding a zinc transporter (ΔzrfC) and a transcription factor (ΔzafA) that regulates zinc absorption show a reduced virulence phenotype in murine models of pulmonary aspergillosis [16, 60].

5. Secondary metabolites

5.1 Melanin

Melanin is one of the most important virulence determinants in *A. fumigatus* [61, 62]. This fungus can synthesize two different types of melanin: DHN-melanin attached to the cell wall of conidia and water-soluble extracellular pyomelanin [61]. The metabolic pathway of DHN-melanin production is activated during conidiation and may be involved in multiple mechanisms of adaptation and survival in harsh environments. DHN-melanin protects against UV rays and desiccation and neutralizes free radicals [63, 64]. In addition, it protects the fungus against the innate immune response, mainly affecting the activity of macrophages, phagocytosis, and acidification of phagosomes and phagolysosomes, which facilitates the colonization and persistence of conidia during infection. It is thought that the production of pyomelanin is a mechanism that protects germinating hyphae when DHN-melanin has disappeared. DHN-melanin is closely associated with the adhesion of molecules such as hydrophobins that form a rod layer that provides conidial hydrophobicity, physical resistance, and immune inertia to *A. fumigatus* against the host's immune system [65]. DHN-melanin is also known to bind to antimicrobial peptides, reduce the efficacy of antifungal drugs, and prevent intracellular destruction of conidia by reducing luminal acidification and resisting phagolysosomal degradation [26].

DHN-melanin biosynthesis in *A. fumigatus* is a polyketide-based pigment synthesis consisting of six genes: *pksP/alb1*, *ayg1*, *arp1*, *arp2*, *abr1*, and *abr2*, expressed during conidiation [66]. *A. fumigatus* mutants lacking any of these genes related to melanin biosynthesis have shown melanin-deficient phenotypes and decreased virulence in *Galleria mellonella* [67]. Since melanin provides cell wall stability and structural rigidity [5], it has been suggested that in mutant strains, the absence of melanin causes a modification of the fungal cell wall, which in turn triggers an increased immune response in the larvae.

5.2 Gliotoxin

Gliotoxin (GT) is a hydrophobic metabolite secreted by *A. fumigatus*, which belongs to the class of epipolythiodioxopiperazine compounds characterized by a quinoid fraction and a disulfide bridge across the piperazine ring, which is essential for its toxicity [11, 68–70]. GT is a recognized virulence factor in this fungus. Its functions are to inhibit the phagocytic activity of macrophages and the response to oxidative stress, decrease the cytotoxic activity of T cells, and prevent the apoptosis induction of host cells [64, 71]. The biological activity of GT is based on an internal disulfide bridge that can bind and inactivate proteins through sulfur: thiol exchange [4]. The transcription factor mtfA regulates GT biosynthesis through the *gliZ* and *gliP* genes. Reeves et al. [72] showed in the *G. mellonella* model that there is a positive correlation between GT production and the pathogenicity of *A. fumigatus*, i.e., isolates with high GT production were lethal to the larva. There is also clinical evidence of GT involvement as a virulence factor. For example, GT has been detected in the lung and the serum of cancer patients suffering from invasive aspergillosis. It should be noted that this finding was corroborated in a murine model. Other evidence is that more than 90% of *A. fumigatus* strains isolated from cancer and invasive aspergillosis patients produce GT, and GT has been seen to occur much faster at 37°C under high oxygen levels, that is, under conditions similar to the host lung environment [11].

5.3 Galactosaminogalactan

Galactosaminogalactan (GAG) is a specific carbohydrate polymer consisting of galactose bound to α-1,4, N-acetyl galactosamine (GalNAc), and galactosamine (GalN), which is expressed and secreted by actively growing hyphae of *A. fumigatus* [10]. After being secreted by hyphae, GAG binds to the surface of the hyphae themselves, generating a polysaccharide sheath that covers the growing fungus and forms an extracellular matrix between the hyphae [2]. Then, it can wrap cell wall polysaccharides such as β-glucans from innate immune detection and repel cationic molecules such as antimicrobial peptides associated with neutrophil extracellular traps [32, 73]. As a cationic exopolysaccharide located within the extracellular matrix, GAG is an adhesin that mediates binding to anionic surfaces, such as human cells, macromolecules, and plastic and supports biofilm formation [32, 74]. Other vital functions of hyphae-secreted GAG are mediating neutrophil apoptosis and resistance to neutrophil extracellular traps (NETs), modulating host immune responses through platelet activation, inducing secretion of IL-1 receptor antagonists, and inflammasome activation [31, 73, 75]. Therefore, GAG is critical for host damage and fungal virulence, as studies in mouse models of invasive pulmonary aspergillosis (IPA) have shown, where GAG-deficient strains exhibit reduced adherence to lung epithelial cells, do not form biofilms, and are less virulent [7, 32]. The multiple and important functions of GAG in virulence have prompted the study of the mechanisms involved in its biosynthesis. To date, it has been seen that GAG synthesis is initiated intracellularly with the interconversion of UDP-N-acetylglucosamine to UDP-N-acetylgalactosamine and UDP-glucose to UDP-galactose by the bifunctional UDP-glucose-4-epimerase (Uge3) [76]. The polymerization and extracellular export of the GAG macromolecule are mediated by a transmembrane glycosyltransferase (Gtb3) [74]. The release of fully acetylated GAG is mediated by a glycoside hydrolase (Sph3) anchored to the membrane that retains endo-α-1,4-N-acetylgalactosaminidase activity [77]. Upon crossing the cell wall, acetylated GAG is processed by a secreted carbohydrate esterase (Agd3), making GAG cationic and biologically active [78, 79].

5.4 Fumagillin

Fumagillin (FM) is a mycotoxin produced by *A. fumigatus* during hyphae development. FM is considered an important virulence factor that inhibits angiogenesis; that is, it reduces the proliferation of endothelial cells for the formation of blood vessels, preventing the infiltration of host immune cells to the infection site [80]. It has been reported that this toxin is produced during the first 30–72 h of invasive aspergillosis, which, together with gliotoxin (GT), helps the fungus evade the immune response and favors the spread and invasion of hyphae, damaging the epithelial layer [11, 81]. Some studies in the *G. mellonella* model have reported that FM exhibits potent neutrophil inhibitory activity, destabilizing the proper immune response to infections. Inhibition of hemocyte phagocytosis is also observed, allowing the fungus to grow in the larva [64, 82]. In other studies, FM has been administered to larvae prior to inoculation with *A. fumigatus*, finding that it increases susceptibility to infection [83, 84]. Therefore, understanding the mechanism of action of FM may open paths to finding new therapies for treating invasive aspergillosis [85].

5.5 Alkaloids

Ergot alkaloids are metabolites produced by *A. fumigatus*. The festuclavine and fumigaclavine alkaloids A, B and C are present in or on the conidia of *A. fumigatus* [86]. Fumigaclavine C inhibits the production of tumor necrosis factor α in human macrophages and reduces the expression of several other inflammatory cytokines in mice [87]. In addition, several in vivo studies in the *G. mellonella* model show that this type of specialized alkaloid is involved in fungal pathogenicity [64]. The most relevant evidence has been that the mutant strains of *A. fumigatus*, generated by the alteration of the genes involved in the biosynthesis of fumigaclavine, both *PesL* and *pes1* and *dmaW*, showed a hypovirulent phenotype in *G. mellonella* due to a deficient production of fumigaclavine C [64, 88, 89].

6. Fungal development

From the two *A. fumigatus* morphotypes, the hyphal morphotype predominates during invasive pulmonary aspergillosis, while the conidial morphotype is rarely observed [90]. It has been reported that during *in vitro* growth, conidia production is associated with a reduction in hyphae growth [91]. Therefore, the forced induction of conidiation during infection could decrease virulence by suppressing the growth of hyphae and invasion. The regulation of conidiation is determined by the transcription factor BrlA. This factor is sufficient to activate the conidiation pathway in *A. fumigatus*, inhibit in vitro vegetative growth, and reduce the virulence of invasive *Aspergillus* infection in vivo models. Likewise, it has been observed that ΔbrlA mutants of *A. fumigatus* do not produce conidia but hyphae and exhibit greater virulence. However, the effects of *brlA* overexpression in *A. fumigatus* are unknown [91].

On the other hand, there are studies in the *G. mellonella* model that have shown some proteins (flbA, gprK, rgsA, rax1, rgsC, and rgsD) of the G-protein signaling regulate, positively or negatively, the production of virulence factors in *A. fumigatus*, such as GT and melanin [64]. In addition, it has been observed that fungal isolates with mutant protein ΔrgsD increase conidiation, stress response, and the production of GT and melanin and, therefore, virulence in the larva. However, with *ΔrgsC* and *ΔgprK* mutants, the conidiation, growth, and tolerance to H_2O_2 and GT production are reduced, the cell wall is modified, and virulence is reduced [64, 88, 92]. In the same way, it has been observed that other GTPase proteins (srgA A, srgA B, srgA C) participate in the development and filamentation of fungi. For example, in mutant ΔsrgA isolates, the growth rate, aberrant conidiation, and virulence are reduced [93].

A. fumigatus has five septins or proteins with GTPase function (aspA, aspB, aspC, aspD, and aspE) involved in regulating critical cellular processes, such as septation. It has been observed that in *G. mellonella* mutant isolates ΔaspA, ΔaspB, and ΔaspC are hypervirulent, and the conidiation is reduced without alteration in the growth rate, except in the case of mutants′ ΔaspB in whom the growth rate is reduced [94].

7. Calcineurin

Calcineurin is an essential virulence factor in *A. fumigatus*. Calcineurin is a specific serine/threonine protein phosphatase that is a heterodimer consisting of a catalytic subunit, CnA, and a regulatory subunit, CnB. It is activated in the presence of calcium and calmodulin. Calcineurin signaling in *A. fumigatus* and other fungal pathogens is highly conserved and leads to the activation of virulence genes and proteins essential for organism growth at host body temperature, hyphae development, and survival [95]. These essential functions in virulence make calcineurin a target for developing antifungals. For example, FK506 (tacrolimus) is a natural calcineurin inhibitor produced by several *Streptomyces* species with potent antifungal activity [96]. FK506 acts on fungal cells by binding to FK506 binding protein 12 (FKBP12), forming a complex that binds to calcineurin and inhibits it by sterically blocking substrate access

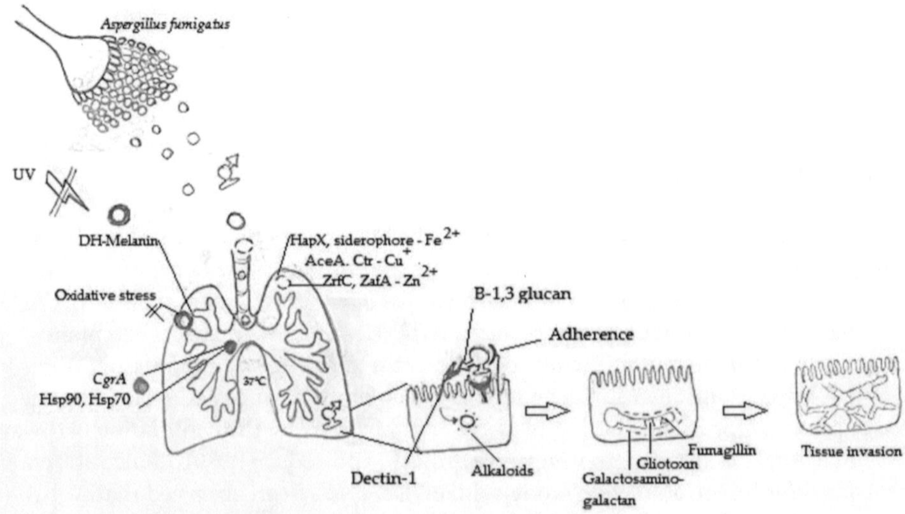

Figure 4.
Mecanisms of virulence in A. fumigatus.

to the active site. FKBP12 belongs to the protein family called immunophilins that bind to immunosuppressive molecules and mediate their activity [97, 98].

So, it is fascinating how this fungus manages to go from harmless to pathogenic as, in addition to the predisposing factors of the human, multiple attributes of the fungus intervene that favor its growth and survival in the host. Among these virulence attributes are thermotolerance, the ability to evade the immune response, some components of the cell wall, the production of secondary metabolites, compliance with nutritional requirements, and the production of melanin, among others (**Figure 4**). Furthermore, some of these virulence attributes are interrelated, making understanding the pathogenesis of aspergillosis more complex.

8. Conclusions

A. fumigatus requires different strategies to infect and cause disease in different hosts and suppress resistance responses by the host. In these strategies, it uses various attributes whose expression is influenced by environmental conditions (nutrient composition and response to host defenses). Some of these virulence attributes are interrelated, which makes it more complex to understand the adaptation mechanisms of both host and fungus to adapt and survive in a hostile environment and produce an infection. However, it is essential to understand these mechanisms to help develop new therapeutic strategies against aspergillosis.

Conflict of interest

The authors declare no conflict of interest.

DOI: http://dx.doi.org/10.5772/intechopen.111778

A. Appendix 1

A.1 *Aspergillus* taxonomy

Aspergillus taxonomy is a complex subject. In 1965, based on phenotypic characteristics, Raper and Fennell [99] classified 150 *Aspergillus* species into 18 groups. In 1985, Gams et al. [100] reclassified the groups into 18 sections as a formal taxonomic status. Currently, the approximately 250 known species within the *Aspergillus* genus are classified into at least 16 sections [101, 102]. This classification is based on a polyphasic analysis that not only includes phenotypic but also molecular studies. The Sections of the current classification are: *Aenei*, *Aspergillus*, *Bispori*, *Candidi*, *Circumdati*, *Clavati*, *Cremei*, *Flavi*, *Flavipedes*, *Fumigati*, *Nidulantes*, *Nigri*, *Restricti*, *Terrei*, *Usti*, *Zonati*. The major species known to cause disease in humans are found in five *Aspergillus* sections: *Fumigati*, *Flavi*, *Nigri*, *Terrei*, and *Nidulante*; however, the *A.* section *Fumigati* is the most frequent. The section *Fumigati* includes 12 species that are pathogenic for humans, several of which have been found to be in the sexual state (*Neosartorya*) (**Table 1**). The species in section *Fumigati* are morphologically indistinguishable from each other, but may present different antifungal susceptibility profiles.

Due to taxonomic complexity, in this chapter we refer simply to *A. fumigatus*.

Species	
Anamorph	**Teleomorph**
A. fumigatus	*Neosartorya fumigata*
Aspergillus udagawae	*Neosartorya udagawae*
Aspergillus pseudofischeri	*Neosartorya pseudofischeri*
Aspergillus lentulus	Unknown
Aspergillus felis	Unknown
Aspergillus hiratsukae	*Neosartorya hiratsukae*
Aspergillus fischeranus	*Neosartorya fischeri*
Aspergillus viridinutans	Unknown
Aspergillus fumisynnematus	Unknown
Aspergillus fumigatiaffinis	Unknown
Aspergillus novofumigatus	Unknown
Aspergillus laciniosa	*Neosartorya laciniosa*

Table 1.
Pathogenic species within the Fumigati *section.*

Author details

María Guadalupe Frías-De-León[1]*, Eduardo García-Salazar[1]
and Gustavo Acosta-Altamirano[2]

1 Unit Research, High Specialty Regional Hospital of Ixtapaluca, Ixtapaluca, Mexico

2 High Specialty Regional Hospital of Ixtapaluca, Ixtapaluca, Mexico

*Address all correspondence to: magpefrias@gmail.com

References

[1] Rokas A. *Aspergillus*. Current Biology 2013;23(5):R187–R188. DOI: 10.1016/j.cub.2013.01.021

[2] Lim JY, Kim YJ, Woo SA, Jeong JW, Lee YR, Kim CH, et al. The LAMMER kinase, LkhA, affects *Aspergillus fumigatus* pathogenicity by modulating reproduction and biosynthesis of cell wall PAMPs. Frontiers in Cellular and Infection Microbiology. 2021;**11**:756206. DOI: 10.3389/fcimb.2021.756206

[3] Dyer PS, Paoletti M. Reproduction in *Aspergillus fumigatus*: Sexuality in a supposedly asexual species? Medical Mycology. 2005;**43**(1):S7-S14. DOI: 10.1080/13693780400029015

[4] Dagenais TR, Keller NP. Pathogenesis of *Aspergillus fumigatus* in invasive aspergillosis. Clinical Microbiology Reviews. 2009;**22**:447-465. DOI: 10.1128/CMR.00055-08

[5] Hoda S, Vermani M, Joshi RK, Shankar J, Vijayaraghavan P. Anti-melanogenic activity of *Myristica fragrans* extract against *Aspergillus fumigatus* using phenotypic based screening. BMC Complementary Medicine and Therapies. 2020a;**20**:67. DOI: 10.1186/s12906-020-2859-z

[6] McCormick A, Loefer J, Ebel F. *Aspergillus fumigatus*: Contours of an opportunistic human pathogen. Cellular Microbiology. 2010;**12**:1535-1543. DOI: 10.1111/j.1462-5822.2010.01517.x

[7] Rahman S, van Rhijn N, Papastamoulis P, Thomson DD, Carter Z, Fortune-Grant R, et al. Distinct cohorts of *Aspergillus fumigatus* transcription factors are required for epithelial damage occurring via contact- or soluble effector-mediated mechanisms. Frontiers in Cellular and Infection Microbiology. 2022;**12**:907519. DOI: 10.3389/fcimb.2022.907519

[8] Gregg KS, Kauffman CA. Invasive aspergillosis: Epidemiology, clinical aspects, and treatment. Seminars in Respiratory and Critical Care Medicine. 2015;**36**:662-672. DOI: 10.1055/s-0035-1562893

[9] Kosmidis C, Denning D. The clinical spectrum of pulmonary aspergillosis. Thorax. 2015;**70**(3):270-277. DOI: 10.1136/thoraxjnl-2014-206291

[10] van de Veerdonk FL, Gresnigt MS, Romani L, Netea MG, Latge JP. *Aspergillus fumigatus* morphology and dynamic host interactions. Nature Reviews. Microbiology. 2017;**15**:661-674. DOI: 10.1038/nrmicro.2017.90

[11] Gayathri L, Akbarsha MA, Ruckmani K. *In vitro* study on aspects of molecular mechanisms underlying invasive aspergillosis caused by gliotoxin and fumagillin, alone and in combination. Scientific Reports. 2020;**10**(1):14473. DOI: 10.1038/s41598-020-71367-2

[12] Steenwyk JL, Mead ME, de Castro PA, Valero C, Damasio A, Dos Santos RAC, et al. Genomic and phenotypic analysis of COVID-19-associated pulmonary aspergillosis isolates of *Aspergillus fumigatus*. Microbiology Spectrum. 2021;**9**(1):e0001021. DOI: 10.1128/Spectrum.00010-21

[13] Hof H, Kupfahl C. Gliotoxin in *Aspergillus fumigatus*: An example that mycotoxins are potential virulence factors. Mycotoxin Research. 2009;**25**(3):123-131. DOI: 10.1007/s12550-009-0020-4

[14] Ghazaei C. Molecular insights into pathogenesis and infection with *Aspergillus fumigatus*. The Malaysian Journal of Medical Sciences. 2017;**24**(1):10-20. DOI: 10.21315/mjms2017.24.1.2

[15] Alonso VA, Velasco Manini MA, Pena GA, Cavaglieri LR. First report on fumagillin production by *Aspergillus fumigatus sensu stricto* gliotoxigenic strains recovered from raw cow milk and clinical samples in Argentina. Revista Argentina de Microbiología. 2022;**54**:243-246. DOI: 10.1016/j.ram.2022.03.003

[16] Earle K, Valero C, Conn DP, Vere G, Cook PC, Bromley MJ, et al. Pathogenicity and virulence of *Aspergillus fumigatus*. Virulence. 2023;**14**:1. DOI: 10.1080/21505594.2023.2172264

[17] Hoda S, Gupta L, Shankar J, Gupta AK, Vijayaraghavan P. Cis-9-Hexadecenal, a natural compound targeting cell wall organization, critical growth factor, and virulence of *Aspergillus fumigatus*. ACS Omega. 2020b;**5**:10077-10088. DOI: 10.1021/acsomega.0c00615

[18] Bhabhra R, Askew DS. Thermotolerance and virulence of *Aspergillus fumigatus*: Role of the fungal nucleolus. Medical Mycology. 2005;**43**(Suppl 1):S87-S93. DOI: 10.1080/13693780400029486

[19] Ryckeboer J, Mergaert J, Coosemans J, Deprins K, Swings J. Microbiological aspects of biowaste during composting in a monitored compost bin. Journal of Applied Microbiology. 2003;**94**:127-137. DOI: 10.1046/j.1365-2672.2003.01800.x

[20] Horianopoulos LC, Kronstad JW. Chaperone networks in fungal pathogens of humans. Journal of Fungi (Basel, Switzerland). 2021;7:209. DOI: 10.3390/jof7030209

[21] Chang YC, Tsai HF, Karos M, Kwon-Chung KJ. THTA, a thermotolerance gene of *Aspergillus fumigatus*. Fungal Genetics and Biology. 2004;**41**:888-896. DOI: 10.1016/j.fgb.2004.06.004

[22] Zhou H, Hu H, Zhang L, Li R, Ouyang H, Ming J, et al. O-Mannosyltransferase 1 in *Aspergillus fumigatus* (AfPmt1p) is crucial for cell wall integrity and conidium morphology, especially at an elevated temperature. Eukaryotic Cell. 2007;**6**:2260-2268. DOI: 10.1128/EC.00261-07

[23] Bhabhra R, Miley MD, Mylonakis E, Boettner D, Fortwendel J, Panepinto JC, et al. Disruption of the *Aspergillus fumigatus* gene encoding nucleolar protein CgrA impairs thermotolerant growth and reduces virulence. Infection and Immunity. 2004;**72**:4731-4740. DOI: 10.1128/IAI.72.8.4731-4740.2004

[24] Lamoth F, Juvvadi PR, Soderblom EJ, Moseley MA, Asfaw YG, Steinbach WJ. Identification of a key lysine residue in heat shock protein 90 required for azole and echinocandin resistance in *Aspergillus fumigatus*. Antimicrobial Agents and Chemotherapy. 2014;**58**(4):1889-1896. DOI: 10.1128/AAC.02286-13

[25] Tiwari S, Shankar J. Hsp70 in fungi: Evolution, function and vaccine candidate. In: HSP70 in Human Diseases and Disorders. Cham, Switzerland: Springer; 2018. pp. 381-400

[26] Croft CA, Culibrk L, Moore MM, Tebbutt SJ. Interactions of *Aspergillus fumigatus* conidia with airway epithelial cells: A critical review. Frontiers in Microbiology. 2016;7:472. DOI: 10.3389/fmicb.2016.00472

[27] Sheppard DC. Molecular mechanism of *Aspergillus fumigatus* adherence to host constituents. Current Opinion

in Microbiology. 2011;**14**(4):375-379. DOI: 10.1016/j.mib.2011.07.006

[28] Rementeria A, López-Molina N, Ludwig A, Vivanco AB, Bikandi J, Pontón J, et al. Genes and molecules involved in *Aspergillus fumigatus* virulence. Revista Iberoamericana de Micología. 2005;**22**(1):1-23. DOI: 10.1016/s1130-1406(05)70001-2

[29] García-Vidal C, Salavert LM. Inmunopatología de las micosis invasivas por hongos filamentosos. Revista Iberoamericana de Micología. 2014;**31**(4):219-228. DOI: 10.1016/j.riam.2014.09.001

[30] Steele C, Rapaka RR, Metz A, Pop SM, Williams DL, Gordon S, et al. The beta-glucan receptor dectin-1 recognizes specific morphologies of *Aspergillus fumigatus*. PLoS Pathogens. 2005;**1**(4):e42. DOI: 10.1371/journal.ppat.0010042

[31] Speth C, Rambach G, Lass-Flörl C, Howell PL, Sheppard DC. Galactosaminogalactan (GAG) and its multiple roles in *Aspergillus* pathogenesis. Virulence. 2019;**10**(1):976-983. DOI: 10.1080/21505594.2019.1568174

[32] Gravelat FN, Beauvais A, Liu H, Lee MJ, Snarr BD, Chen D, et al. *Aspergillus* galactosaminogalactan mediates adherence to host constituents and conceals hyphal b-glucan from the immune system. PLoS Pathogens. 2013;**9**(8):e1003575. DOI: 10.1371/journal.ppat.1003575

[33] Zhang S, Chen Y, Ma Z, et al. PtaB, a lim-domain binding protein in *Aspergillus fumigatus* regulates biofilm formation and conidiation through distinct pathways. Cellular Microbiology. 2018;**20**:e12799. DOI: 10.1111/cmi.12799

[34] Beaussart A, El-Kirat-Chatel S, Fontaine T, Latgé JP, Dufrêne YF. Nanoscale biophysical properties of the cell surface galactosaminogalactan from the fungal pathogen *Aspergillus fumigatus*. Nanoscale. 2015;**7**(36):14996-15004. DOI: 10.1039/c5nr04399a

[35] Henriet SS, van de Sande WW, Lee MJ, Simonetti E, Momany M, Verweij PE, et al. Decreased cell wall galactosaminogalactan in *Aspergillus nidulans* mediates dysregulated inflammation in the chronic granulomatous disease host. Journal of Interferon & Cytokine Research. 2016;**36**(8):488-498. DOI: 10.1089/jir.2015.0095

[36] Garcia-Vidal C, Carratalà J. Patogenia de la infección fúngica invasora. Enfermedades Infecciosas y Microbiología Clínica. 2012;**30**(3):151-158. DOI: 10.1016/j.eimc.2011.09.011

[37] Antsotegi-Uskola M, Markina-Iñarrairaegui A, Ugalde U. New insights into copper homeostasis in filamentous fungi. International Microbiology. 2020;**23**:65-73. DOI: 10.1007/s10123-019-00081-5

[38] Balamurugan K, Schaffner W. Copper homeostasis in eukaryotes: Teetering on a tightrope. Biochimica et Biophysica Acta. 2006;**1763**(7):737-746. DOI: 10.1016/j.bbamcr.2006.05.001

[39] Gerwien F, Skrahina V, Kasper L, Hube B, Brunke S. Metals in fungal virulence. FEMS Microbiology Reviews. 2018;**42**(1):fux050. DOI: 10.1093/femsre/fux050

[40] Puig S, Lee J, Lau M, Thiele DJ. Biochemical and genetic analyses of yeast and human high affinity copper transporters suggest a conserved mechanism for copper uptake. The Journal of Biological Chemistry. 2002;**277**(29):26021-26030. DOI: 10.1074/jbc.M202547200

[41] Petris MJ. The SLC31 (Ctr) copper transporter family. Pflügers Archiv. 2004;**447**:752-755. DOI: 10.1007/s00424-003-1092-1

[42] Ding C, Festa RA, Sun TS, Wang ZY. Iron and copper as virulence modulators in human fungal pathogens. Molecular Microbiology. 2014;**93**(1):10-23. DOI: 10.1111/mmi.12653

[43] Jiang J, Nadas IA, Kim MA, Franz KJ. A Mets motif peptide found in copper transport proteins selectively binds Cu(I) with methionineonly coordination. Inorganic Chemistry. 2005;**44**(26):9787-9794. DOI: 10.1021/ic051180m

[44] De Feo CJ, Aller SG, Siluvai GS, Blackburn NJ, Unger VM. Three-dimensional structure of the human copper transporter hCTR1. Proceedings of the National Academy of Sciences of the United States of America. 2009;**106**:4237-4242. DOI: 10.1073/pnas.0810286106

[45] Cai Z, Du W, Zhang Z, Guan L, Zeng Q, Chai Y, et al. The *Aspergillus fumigatus* transcription factor AceA is involved not only in Cu but also in Zn detoxification through regulating transporters CrpA and ZrcA. Cellular Microbiology. 2018;**20**:e12864. DOI: 10.1111/cmi.12864

[46] Garcia-Santamarina S, Thiele DJ. Copper at the fungal pathogenhost axis. The Journal of Biological Chemistry. 2015;**290**:18945-18953. DOI: 10.1074/jbc.R115.649129

[47] Upadhyay S, Torres G, Lin X. Laccases involved in 1,8-dihydroxynaphthalene melanin biosynthesis in *Aspergillus fumigatus* are regulated by developmental factors and copper homeostasis. Eukaryotic Cell. 2013;**12**(12):1641-1652. DOI: 10.1128/EC.00217-13

[48] Pihet M, Vandeputte P, Tronchin G, Renier G, Saulnier P, Georgeault S, et al. Melanin is an essential component for the integrity of the cell wall of *Aspergillus fumigatus* conidia. BMC Microbiology. 2009;**9**:177. DOI: 10.1186/1471-2180-9-177

[49] Park YS, Lian H, Chang M, Kang CM, Yun CW. Identification of high-affinity copper transporters in *Aspergillus fumigatus*. Fungal Genetics and Biology. 2014;**73**:29-38. DOI: 10.1016/j.fgb.2014.09.008

[50] Kurucz V, Krüger T, Antal K, Dietl AM, Haas H, Pócsi I, et al. Additional oxidative stress reroutes the global response of *Aspergillus fumigatus* to iron depletion. BMC Genomics. 2018;**19**(1):357. DOI: 10.1186/s12864-018-4730-x

[51] Haas H. Iron: A key nexus in the virulence of *Aspergillus fumigatus*. Frontiers in Microbiology. 2012;**3**:28. DOI: 10.3389/fmicb.2012.00028

[52] Matthaiou EI, Sass G, Stevens DA, Hsu JL. Iron: An essential nutrient for *Aspergillus fumigatus* and a fulcrum for pathogenesis. Current Opinion in Infectious Diseases. 2018;**31**(6):506-511. DOI: 10.1097/QCO.0000000000000487

[53] Cassat JE, Skaar EP. Iron in infection and immunity. Cell Host & Microbe. 2013;**13**:509-519. DOI: 10.1016/j.chom.2013.04.010

[54] Gsaller F, Hortschansky P, Beattie SR, Klammer V, Tuppatsch K, Lechner BE, et al. The Janus transcription factor HapX controls fungal adaptation to both iron starvation and iron excess. The EMBO Journal. 2014;**33**(19):2261-2276. DOI: 10.15252/embj.201489468

[55] Schrettl M, Kim HS, Eisendle M, Kragl C, Nierman WC, Heinekamp T,

et al. SreA-mediated iron regulation in *Aspergillus fumigatus*. Molecular Microbiology. 2008;**70**:27-43. DOI: 10.1111/j.1365-2958.2008.06376.x

[56] Gonçalves SM, Lagrou K, Duarte-Oliveira C, Maertens JA, Cunha C, Carvalho A. The microbiome-metabolome crosstalk in the pathogenesis of respiratory fungal diseases. Virulence. 2017;**8**(6):673-684. DOI: 10.1080/21505594.2016.1257458

[57] Skriba A, Pluhacek T, Palyzova A, Novy Z, Lemr K, Hajduch M, et al. Early and non-invasive diagnosis of aspergillosis revealed by infection kinetics monitored in a rat model. Frontiers in Microbiology. 2018;**9**:2356. DOI: 10.3389/fmicb.2018.02356

[58] Schrettl M, Beckmann N, Varga J, Heinekamp T, Jacobsen ID, Jochl C, et al. HapX-mediated adaption to iron starvation is crucial for virulence of *Aspergillus fumigatus*. PLoS Pathogens. 2010;**6**:e1001124. DOI: 10.1371/journal.ppat.1001124

[59] Perez-Cuesta U, Guruceaga X, Cendon-Sanchez S, Pelegri-Martinez E, Hernando FL, Ramirez-Garcia A, et al. Nitrogen, iron and zinc acquisition: Key nutrients to *Aspergillus fumigatus* virulence. Journal of Fungi (Basel, Switzerland). 2021;7(7):518. DOI: 10.3390/jof7070518

[60] Amich J, Vicente Franqueira R, Leal F, Calera JA. *Aspergillus fumigatus* survival in alkaline and extreme zinc-limiting environments relies on the induction of a zinc homeostasis system encoded by the zrfc and aspf2 genes. Eukaryotic Cell. 2010;**9**(3):424-437. DOI: 10.1128/EC.00348-09

[61] Perez-Cuesta U, Aparicio-Fernandez L, Guruceaga X, Martin-Souto L, Abad-Diaz-de-Cerio A, Antoran A, et al. Melanin and pyomelanin in *Aspergillus fumigatus*: From its genetics to host interaction. International Microbiology. 2020;**23**(1):55-63. DOI: 10.1007/s10123-019-00078-0

[62] Paulussen C, Hallsworth JE, Álvarez-Pérez S, Nierman WC, Hamill PG, Blain D, et al. Ecology of aspergillosis: Insights into the pathogenic potency of *Aspergillus fumigatus* and some other *aspergillus* species. Microbial Biotechnology. 2017;**10**(2):296-322. DOI: 10.1111/1751-7915.12367

[63] Shankar J, Tiwari S, Shishodia SK, Gangwar M, Hoda S, Thakur R, et al. Molecular insights into development and virulence determinants of *Aspergilli*: A proteomic perspective. Frontiers in Cellular and Infection Microbiology. 2018;**8**:180. DOI: 10.3389/fcimb.2018.00180

[64] Durieux MF, Melloul É, Jemel S, Roisin L, Dardé ML, Guillot J, et al. *Galleria mellonella* as a screening tool to study virulence factors of *Aspergillus fumigatus*. Virulence. 2021;**12**(1):818-834. DOI: 10.1080/21505594.2021.1893945

[65] Bayry J, Beaussart A, Dufrene YF, Sharma M, Bansal K, Kniemeyer O, et al. Surface structure characterization of *Aspergillus fumigatus* conidia mutated in the melanin synthesis pathway and their human cellular immune response. Infection and Immunity. 2014;**82**:3141-3153. DOI: 10.1128/IAI.01726-14

[66] Jacob S, Grötsch T, Foster AJ, Schüffler A, Rieger PH, Sandjo LP, et al. Unravelling the biosynthesis of pyriculol in the rice blast fungus *Magnaporthe oryzae*. Microbiology. 2017;**163**(4):541-553. DOI: 10.1099/mic.0.000396

[67] Fallon JP, Troy N, Kavanagh K. Pre-exposure of *galleria mellonella* larvae

to different doses of *Aspergillus fumigatus* conidia causes differential activation of cellular and humoral immune responses. Virulence. 2011;**2**(5):413-421. DOI: 10.4161/viru.2.5.17811

[68] Dolan SK, O'Keeffe G, Jones GW, Doyle S. Resistance is not futile: Gliotoxin biosynthesis, functionality and utility. Trends in Microbiology. 2015;**23**:419-428. DOI: 10.1016/j.tim.2015.02.005

[69] Scharf DH, Heinekamp T, Brakhage AA. Human and plant fungal pathogens: The role of secondary metabolites. PLoS Pathogens. 2014;**10**:e1003859. DOI: 10.1371/journal.ppat.1003859

[70] Scharf DH, Brakhage AA, Mukherjee PK. Gliotoxin - bane or boon? Environmental Microbiology. 2016;**18**:1096-1109. DOI: 10.1111/1462-2920.13080

[71] Smith TD, Calvo AM. The mtfA transcription factor gene controls morphogenesis, gliotoxin production, and virulence in the opportunistic human pathogen *Aspergillus fumigatus*. Eukaryotic Cell. 2014;**13**(6):766-775. DOI: 10.1128/EC.00075-14

[72] Reeves EP, Messina CGM, Doyle S, Kavanagh K. Correlation between gliotoxin production and virulence of *Aspergillus fumigatus* in *galleria mellonella*. Mycopathologia. 2004;**158**(1):73-79. DOI: 10.1023/b:myco.0000038434.55764.16

[73] Lee MJ, Liu H, Barker BM, Snarr BD, Gravelat FN, Al Abdallah Q, et al. The fungal exopolysaccharide galactosaminogalactan mediates virulence by enhancing resistance to neutrophil extracellular traps. PLoS Pathogens. 2015;**11**(10):e1005187. DOI: 10.1371/journal.ppat.1005187

[74] Le Mauff F. Exopolysaccharides and biofilms. Current Topics in Microbiology and Immunology. 2020;**425**:225-254. DOI: 10.1007/82_2020_199

[75] Briard B, Fontaine T, Samir P, Place DE, Muszkieta L, Malireddi RKS, et al. Galactosaminogalactan activates the inflammasome to provide host protection. Nature. 2020;**588**(7839):688-692. DOI: 10.1038/s41586-020-2996-z

[76] Lee MJ, Gravelat FN, Cerone RP, Baptista SD, Campoli PV, Choe SI, et al. Overlapping and distinct roles of *Aspergillus fumigatus* UDP-glucose 4-epimerases in galactose metabolism and the synthesis of galactose-containing cell wall polysaccharides. The Journal of Biological Chemistry. 2014;**289**(3):1243-1256. DOI: 10.1074/jbc.M113.522516

[77] Le Mauff F, Bamford NC, Alnabelseya N, Zhang Y, Baker P, Robinson H, et al. Molecular mechanism of *Aspergillus fumigatus* biofilm disruption by fungal and bacterial glycoside hydrolases. The Journal of Biological Chemistry. 2019;**294**(28):10760-10772. DOI: 10.1074/jbc.RA119.008511

[78] Bamford NC, Le Mauff F, Subramanian AS, Yip P, Millan C, Zhang Y, et al. Ega3 from the fungal pathogen *Aspergillus fumigatus* is an endo-alpha-1,4-galactosaminidase that disrupts microbialbiofilms. The Journal of Biological Chemistry. 2019;**294**(37):13833-13849. DOI: 10.1074/jbc.RA119.009910

[79] Bamford NC, Le Mauff F, Van Loon JC, Ostapska H, Snarr BD, Zhang Y, et al. Structural and biochemical characterization of the exopolysaccharide deacetylase Agd3 required for *Aspergillus fumigatus* biofilm formation. Nature Communications. 2020;**11**(1):2450. DOI: 10.1038/s41467-020-16144-5

[80] Guruceaga X, Ezpeleta G, Mayayo E, Sueiro-Olivares M, Abad-Diaz-De-Cerio A, Aguirre Urízar JM, et al. A possible role for fumagillin in cellular damage during host infection by *Aspergillus fumigatus*. Virulence. 2018;**9**(1):1548-1561. DOI: 10.1080/21505594.2018.1526528

[81] Fallon JP, Reeves EP, Kavanagh K. Inhibition of neutrophil function following exposure to the *Aspergillus fumigatus* toxin fumagillin. Journal of Medical Microbiology. 2010;**59**:625-633. DOI: 10.1099/jmm.0.018192-0

[82] Fallon JP, Reeves EP, The KK. *Aspergillus fumigatus* toxin fumagillin suppresses the immune response of *Galleria mellonella* larvae by inhibiting the action of haemocytes. Microbiology. 2011;**157**(5):1481-1488. DOI: 10.1099/mic.0.043786-0

[83] Jia X, Zhang X, Hu Y, Hu M, Tian S, Han X, et al. Role of actin depolymerizing factor cofilin in *Aspergillus fumigatus* oxidative stress response and pathogenesis. Current Genetics. 2018;**64**(3):619-634. DOI: 10.1007/s00294-017-0777-5

[84] Kim Y, Lee MW, Jun SC, Choi YH, Yu JH, Shin KS. RgsD negatively controls development, toxigenesis, stress response, and virulence in *Aspergillus fumigatus*. Scientific Reports. 2019;**9**:811. DOI: 10.1038/s41598-018-37124-2

[85] Wiesner DL, Klein BS. Lung epithelium: Barrier immunity to inhaled fungi and driver of fungal-associated allergic asthma. Current Opinion in Microbiology. 2017;**40**:8-13. DOI: 10.1016/j.mib.2017.10.007

[86] Robinson SL, Panaccione DG. Diversification of ergot alkaloids in natural and modified fungi. Toxins. 2015;7(1):201-218. DOI: 10.3390/toxins7010201

[87] Du RH, Li EG, Cao Y, Song YC, Tan RX. Fumigaclavine C inhibits tumor necrosis factor α production via suppression of toll-like receptor 4 and nuclear factor κB activation in macrophages. Life Sciences. 2011;**89**(7-8):235-240. DOI: 10.1016/j.lfs.2011.06.015

[88] Reeves EP, Reiber K, Neville C, Scheibner O, Kavanagh K, Doyle S. A nonribosomal peptide synthetase (Pes1) confers protection against oxidative stress in *Aspergillus fumigatus*. The FEBS Journal. 2006;**273**(13):3038-3053. DOI: 10.1111/j.1742-4658.2006.05315.x

[89] O'Hanlon KA, Gallagher L, Schrettl M, Jöchl C, Kavanagh K, Larsen TO, et al. Nonribosomal peptide synthetase genes pesL and pes1 are essential for Fumigaclavine C production in *Aspergillus fumigatus*. Applied and Environmental Microbiology. 2012;**78**(9):3166-3176. DOI: 10.1128/AEM.07249-11

[90] Tochigi N, Okubo Y, Ando T, Wakayama M, Shinozaki M, Gocho K, et al. Histopathological implications of *Aspergillus* infection in lung. Mediators of Inflammation. 2013;**2013**:809798. DOI: 10.1155/2013/809798

[91] Mah JH, Yu JH. Upstream and downstream regulation of asexual development in *Aspergillus fumigatus*. Eukaryotic Cell. 2006;**5**:1585-1595. DOI: 10.1128/EC.00192-06

[92] Jung MG, Kim SS, Yu JH, Shin KS. Characterization of gprK encoding a putative hybrid G-protein-coupled receptor in *Aspergillus fumigatus*. PLoS One. 2016;**11**(9):e0161312. DOI: 10.1372/journal.pone.0161312

[93] Lindsey R, Cowden S, Hernández-Rodríguez Y, Momany M. AspC are important for ormal development and limit the emergence of new growth foci in the multicellular fungus *Aspergillus nidulans*. Eukaryotic Cell. 2010;**9**(1):155-163. DOI: 10.1128/EC.00269-09

[94] Gerhards N, Neubauer L, Tudzynski P, Li SM. Biosynthetic pathways of ergot alkaloids. Toxins (Basel). 2014;**6**(12):3281-3295. DOI: 10.3390/toxins6123281

[95] Hoy MJ, Park E, Lee H, Lim WY, Cole DC, DeBouver ND, et al. Structure-guided synthesis of FK506 and FK520 analogs with increased selectivity exhibit *in vivo* therapeutic efficacy against *Cryptococcus*. MBio. 2022;**13**(3):e0104922. DOI: 10.1128/mbio.01049-22

[96] Goranovič D, Blažič M, Magdevska V, Horvat J, Kuščer E, Polak T, et al. FK506 biosynthesis is regulated by two positive regulatory elements in *Streptomyces tsukubaensis*. BMC Microbiology. 2012;**12**:238. DOI: 10.1186/1471-2180-12-238

[97] Lee Y, Lee KT, Lee SJ, Beom JY, Hwangbo A, Jung JA, et al. *In vitro* and *in vivo* assessment of FK506 analogs as novel antifungal drug candidates. Antimicrobial Agents and Chemotherapy. 2018;**62**(11):e01627-e01618. DOI: 10.1128/AAC.01627-18

[98] Wu QB, Zhang XY, Chen XA, Li YQ. Improvement of FK506 production via metabolic engineering-guided combinational strategies in *Streptomyces tsukubaensis*. Microbial Cell Factories. 2021;**20**(1):166. DOI: 10.1186/s12934-021-01660-w

[99] Raper KB, Fennell DI. The Genus *Aspergillus*. Baltimore: Williams and Wilkins; 1965

[100] Gams W, Christensen M, Onions AH, Pitt JI, Samson RA. Infrageneric taxa of *Aspergillus*. In: Samson RA, Pitt J, editors. Advances in *Penicillium* and *Aspergillus* Systematics. New York: Plenum; 1985. pp. 55-62

[101] Houbraken J, Samson RA. Phylogeny of *Penicillium* and the segregation of *Trichocomaceae* into three families. Studies in Mycology. 2011;**70**(1):1-51. DOI: 10.3114/sim.2011.70.01

[102] Krimitzas A, Pyrri I, Kouvelis VN, Kapsanaki-Gotsi E, Typas MA. A phylogenetic analysis of Greek isolates of *Aspergillus* species based on morphology and nuclear and mitochondrial gene sequences. BioMed Research International. 2013;**2013**:260395. DOI: 10.1155/2013/260395

Section 3

Novel Diagnostic Tools

Chapter 5

Nanomaterials-Based Biosensors against *Aspergillus* and Aspergillosis: Control and Diagnostic Perspectives

Xiaodong Guo, Mengke Zhang, Mengzhi Wang, Jiaqi Wang and Marie-Laure Fauconnier

Abstract

Aspergillosis is the name given to the spectrum of diseases caused by the genus *Aspergillus*. Research on aspergillosis has shown a progressive expansion over the past decades, largely due to the rise in the number of immunocompromised individuals who are at risk for the infection. Nanotechnology provides innovative tools in the medicine, diagnosis, and treatment. The unique properties of nanomaterials like small size in the nanoscale have attracted researchers to explore their potential, especially in medical diagnostics. Aptamers, considered as chemical antibody, are short, single-stranded oligonucleotide molecules with high affinity and specificity to interact with target molecules even superior to antibody. Accordingly, development of nanomaterials-based biosensors technology such as immunosensors and aptasensors against *Aspergillus* and Aspergillosis is of great significance and urgency. In this book chapter, we comprehensively introduce and analyze the recent progress of nanomaterials-based biosensors against *Aspergillus* and Aspergillosis. In addition, we reveal the challenges and provide our opinion in future opportunities for such sensing platform development. Ultimately, conclusion and future prospects are highlighted and summarized.

Keywords: nanomaterials, biosensor, *Aspergillus*, Aspergillosis, biomarker

1. Introduction

Biosensors are integrated analytical devices that are capable of transferring the binding events between bioreceptor and target into detectable optical and electric signals. Therefore, bioreceptors, regarded as the recognition elements, play a crucial role for constructing advanced biosensors. Conventional bioreceptors like antibodies have been extensively prepared and developed for immunosensors' establishment. The barriers such as high cost, complicated procedures, and lack of stability limited the wide applications of such sensing approaches [1–3]. Fortunately, nucleic acid aptamers are short and single-stranded oligonucleotides, selected and identified by

SELEX (Systematic Evolution of Ligands Exponential Enrichment) *in vitro* selection process [4, 5]. Aptamer, considered as chemical antibody and even superior to antibody, can recognize the target molecule to form unique 3D configuration with high affinity and selectivity [6, 7]. Accordingly, aptasensors have attracted increasing attention in recent years toward a variety of target molecules including proteins, cells, viruses, bacteria, metal ions, as well as the disease biomarkers [8, 9].

It is worth noting that the current biosensors generally suffer from the concerns such as lack of biocompatibility, low stability, as well as the poor detection sensitivity. Various advanced nanomaterials can be integrated into the sensing systems for signal transduction and improved analytical performance [10–12]. The commonly used nanomaterials for biosensors mainly include fluorophores, quantum dots (QDs), graphene oxide (GO), gold nanoparticles (AuNPs), silver nanoparticles (AgNPs), metal-organic frameworks (MOFs), upconversion nanoparticles (UCNPs), zinc oxide (ZnO) and other semiconducting nanomaterials, and so on [13–15]. Fascinatingly, bioreceptors can be universally designed and modified with these nanomaterials and are available for optical and electrochemical signal transduction strategies, which further are measured by portable detectors such as fluorimeter, naked eyes, strip readout, as well as electrochemical detectors [16, 17]. For instance, in our recent

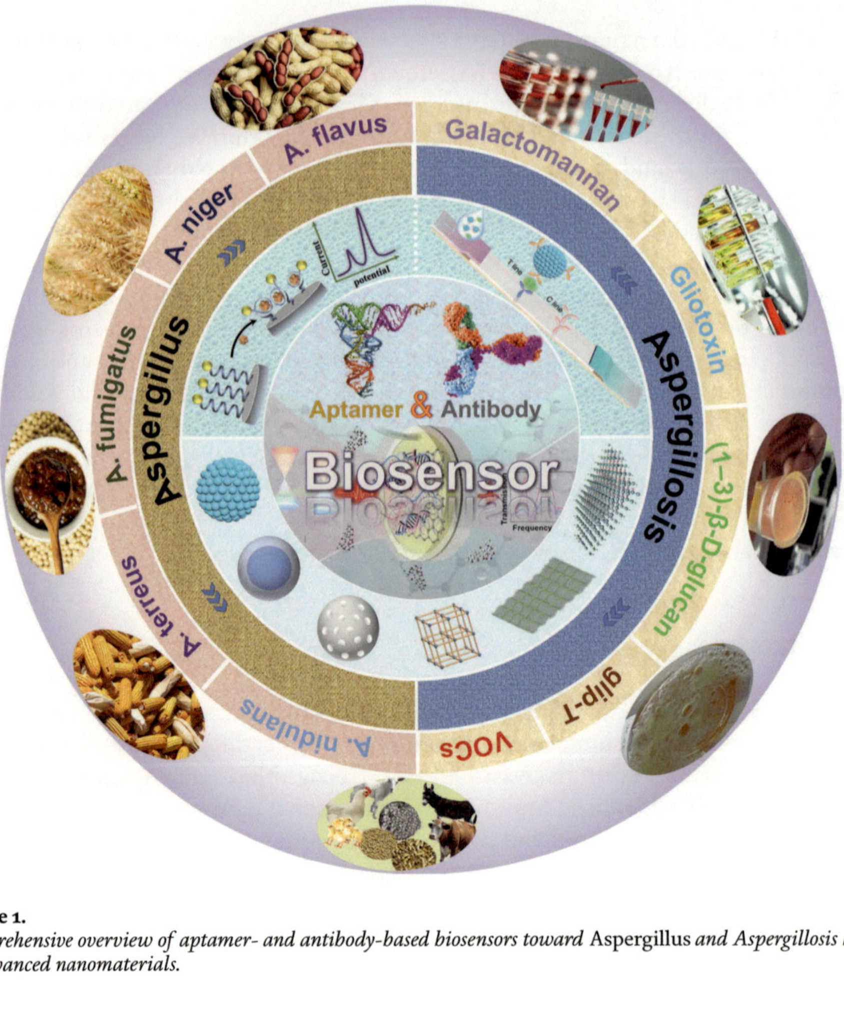

Figure 1.
Comprehensive overview of aptamer- and antibody-based biosensors toward Aspergillus *and Aspergillosis based on advanced nanomaterials.*

study, we have introduced a fluorescent aptasensor toward mycotoxin fumonisin B1 (FB1) based on the aptamer recognition. In this effort, GO was embedded for fluorescent quenching and acted as a protectant of the specific aptamer from nuclease cleavage. The target cycling was then triggered for signal amplification and improved detection sensitivity [9].

At the present time, great advances have been achieved for immunosensors and aptasensors in numerous hazard control. Nevertheless, in this book chapter, we only outlined and highlighted the recent progress toward *Aspergillus* and Aspergillosis (**Figure 1**). *Aspergillus* is an extraordinary fungus that occurs naturally all around the world. Sixteen of the 200 known species of *Aspergillus* are isolated and identified due to their severe hazards to animals and humans. Prolonged exposure to high levels of *Aspergillus* can induce allergic symptoms, toxic symptoms, and infection. In particular, for instance, *Aspergillus fumigatus* is the most common and predominate one that can cause invasive aspergillosis (IA) [18, 19]. Conventional approaches for the detection of *Aspergillus* detection like biopsies of cerebral lesions and extraction of cerebrospinal fluid are usually not appropriate for immunocompromised patients. Therefore, simple, rapid, and accurate analytical techniques for *Aspergillus* and Aspergillosis diagnosis are of great importance for human health.

Immunosensors and aptasensors have witnessed a remarkable progress over the past decade. To the best of our knowledge, the comprehensive discussion on immunosensors and aptasensors toward *Aspergillus* and Aspergillosis diagnosis has not yet been reported. Inspired by this status, we proposed an overview of the biosensor construction and their improved performance (**Figure 1**). Moreover, we highlighted the challenges and new opportunities of advanced biosensors for early diagnosis of Aspergillosis infection.

2. Recent advances of biosensors

In the past three decades, antibody-based immunoassays have been well established from scientists in medical and biotechnology fields, including biosensing, therapy, environment, and so on, and gradually extend to disease diagnosis for various targets and biomarker identification and determination. Classic immunoassays like ELISA (enzyme-linked immunosorbent assays) and LFIA (lateral flow immunoassay) are frequently realized for screening purposes and commercial application in the market [20–22]. In particular, ELISA kit, one of the most widely available products, is suitable for qualitative and quantitative assessment of target biomolecule [23]. The ELISA-based rapid screening methods exhibit desirable sensitivity and selectivity, as well as ease of operation [24, 25]. However, high cost in antibody production and antibody stability issue in complicated environment restrict their extended applications [26, 27]. More importantly, antibody preparation against small molecule remains a rigorous challenge since its non-immunogenic and antibody generation are not significant in the process of animal immune [28–30].

Fortunately, nucleic acid aptamer, considered as "chemical antibody," is short ssDNA or RNA that shows excellent affinity and specificity against its target biomolecule even superior to antibody. Noteworthy, the aptamer selection is performed *in vitro* by SELEX technique instead of the complicated animal experiments *in vivo* [10]. Therefore, compared to antibody, the aptamer possesses obvious advantages like convenient production, low cost, high stability, non-immunogenicity, ease of modification, as well as various targets (proteins, cell, tissue, even small molecules) [31, 32].

Correspondingly, the past decade witnessed a remarkable progress on the development of novel aptasensors against *Aspergillus* and Aspergillosis and mycotoxin contamination. However, it is worth noting that there were few studies focusing on the comparison of analytical performance between antibody and aptamer. Stimulated by this status, we noticed that previous works on aptasensors demonstrated the more excellent analytical performance than that of immunosensors [33, 34]. Therefore, the discovery of specific aptamer and its further research opened up a new horizon for rapid and accurate determination of various hazards with high sensitivity and selectivity.

3. Applications of biosensing strategies for *Aspergillus*

Aspergillus has received global concern due to its hazards on food spoilage and food safety [35]. In particular, *Aspergillus flavus* contamination can produce aflatoxins and cause Aspergillosis [36]. *AflD* gene, a potential biomarker for *Aspergillus flavus* pollution, is a structural gene in aflatoxins gene cluster of *Aspergillus* species like *Aspergillus flavus* [37]. Nevertheless, the analytical strategies for *Aspergillus* control are relatively less reported. Biosensors are thus emerged as advanced techniques for the detection of DNA attributed to their portability, high sensitivity, as well as ease of operation [38]. Sedighi-Khavidak et al. firstly fabricated a novel biosensor toward *aflD* gene analyses of *Aspergillus flavus* based on impedimetric electrochemical signal detection and Au NPs [39]. The Au NPs modified with specific DNA probe were immobilized on the glassy carbon electrode. The presence of target DNA induced the formation of double-stranded DNA *via* hybridization reaction, disrupting the reduction of $[Fe(CN)_6]^{3-}$ and enhancing the electrochemical signal (**Figure 2A**). The electrochemical biosensor exhibited a dynamic response of *aflD* gene that ranged from 1 nM to 10 μM with an LOD of 0.55 nM. Furthermore, the feasibility of this sensing protocol was confirmed for the detection of *aflD* gene in real pistachio samples.

Very recently, Liang et al. have synthesized Cu-anchored PDA (polydopamine) nanomaterials, which exhibited photothermal property and catalytic ability for colorimetric signal [40]. In this regard, the nanomaterials were further embedded into LFIA for multimodal sensing of *Aspergillus flavus* based on a sandwich design (**Figure 2B**). Consequently, compared to traditional colorimetric method, the novel LFIA platform showed a significantly sensitive detection of *Aspergillus flavus* with LODs of 0.45 and 0.22 ng/mL, respectively. The feasibility of the sensing strategy was further investigated to monitor *Aspergillus flavus* in peanut and maize samples. On the other hand, apart from the detection of *Aspergillus flavus*, volatile organic compounds (VOCs), the representative metabolites of *Aspergillus flavus*, could be identified and detected for *Aspergillus flavus* contamination [43]. In order to improve the detection efficiency, Lin et al. incorporated three types of nanomaterials into nanocomposites for colorimetric sensing of VOCs [41]. MOF (metal-organic framework), PSA (poly styrene-co-acrylic acid), and PSN (porous silica nanoparticles) were employed and integrated for improved performance due to their ultra-large surface area, excellent catalytic property, and numerous binding sites (**Figure 2C**). The developed sensing method showed high sensitivity and stability for the detection of VOCs and great promise for wheat mildew monitoring. From another viewpoint, the identification and detection of VOCs is significantly correlated with food spoilage event [44]. Traditional analytical technologies like gas chromatography-ion mobility spectrometry (GC-IMS) [45] and electronic nose [46] require expensive instruments, professional personnel, and complicated procedures. Fortunately, whole-cell biosensor

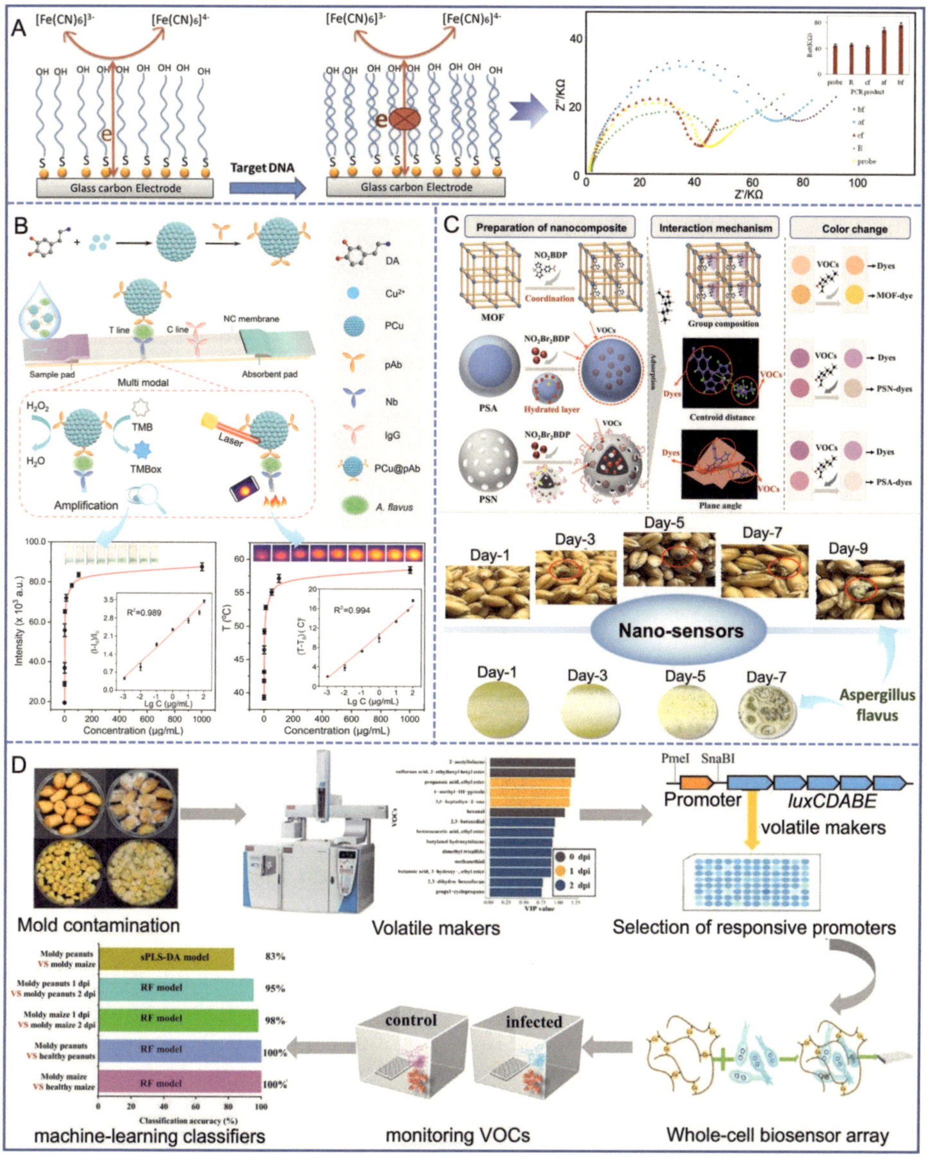

Figure 2.
(A) Schematic representation of impedimetric electrochemical DNA sensor for sensitive detection of aflD *gene of* Aspergillus flavus *based on the Au NPs modification on electrode. The illustration was recreated according to ref. [39]. Copyright 2017: Taylor & Francis Group, LLC. (B) Schematic diagram of lateral flow immunoassay for colorimetric and photothermal detection of* Aspergillus flavus *via the functionalization of Cu-anchored PDA. The illustration was recreated according to ref. [40]. Copyright 2022: Elsevier. (C) Schematic illustration of the colorimetric biosensor toward* Aspergillus flavus *attributed to the analysis of VOCs by integrating nanocomposites. The illustration was recreated according to ref. [41]. Copyright 2022: Elsevier. (D) Mechanism illustration of a novel whole-cell biosensor for identification and monitoring of VOCs and* Aspergillus *contamination by integrating machine-learning models. The illustration was recreated according to ref. [42]. Copyright 2023: Elsevier.*

possesses a great promise for rapid, portable, highly efficient, and sensitive control of VOCs [47]. Herein, Ma et al. developed a novel whole-cell biosensing platform for VOCs finding and *Aspergillus* infection monitoring by engineering machine learning

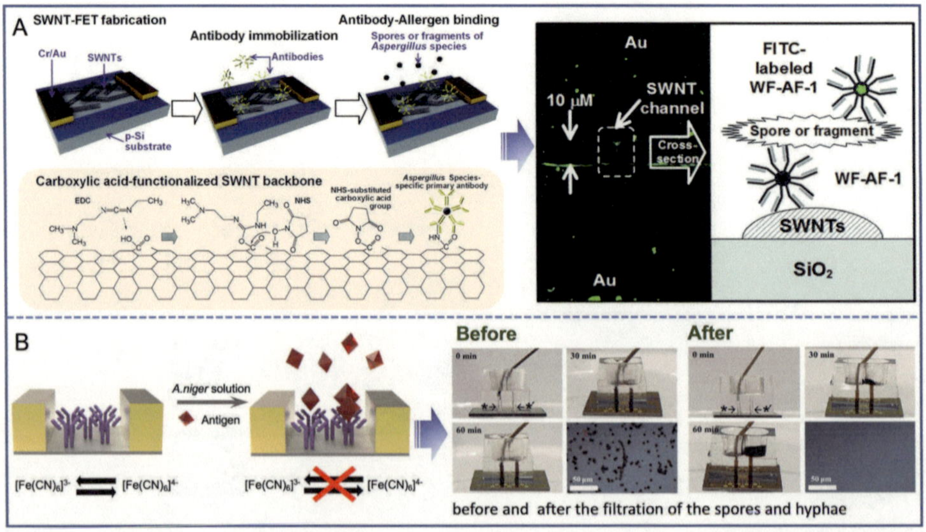

Figure 3.
(A) Schematic illustration of the novel immunosensor for highly sensitive detection of Aspergillus Niger *based on the functionalization of single-walled carbon nanotube (SWNT). The illustration was recreated according to ref. [51]. Copyright 2015: Royal Society of Chemistry. (B) Schematic diagram of electrochemical immunosensor for selective quantification of the* Aspergillus Niger *based on the extracellular proteins monitoring and antibody–antigen reaction. The illustration was recreated according to ref. [52]. Copyright 2021: Elsevier.*

models [42]. Three VOCs' markers were identified in peanut by *Aspergillus flavus* infection and further realized for the validation of various response modes and the construction of biosensor (**Figure 2D**). Moreover, the proposed biosensor coupled with machine-learning models exhibited excellent prediction accuracy in both infected matrices and pre-mold stages. Hence, this novel biosensing strategy opened a new avenue for *Aspergillus* infection prediction and food control.

Conventional *Aspergillus* analytical methods generally involve sampling and culturing of *Aspergillus* spores for immunoassay signal detection, followed by DNA sequencing and quantitative analysis. These protocols suffer from the drawbacks like high cost, complicated procedures, and time-consuming [48–50]. To overcome the barrier, advanced nanomaterials-based biosensors are attracting increasing attention for improved performance. For instance, Jin et al. developed a nanoscale immunosensor for real-time monitoring of *Aspergillus niger via* single-walled carbon nanotube (SWNT) and antigen-antibody recognition (**Figure 3A**) [51]. Encouragingly, the integrating of carbon nanomaterials significantly increased the antibody immobilization sites for enhanced detection signal and high sensitivity at sub-picomolar levels. On the basis of similar antigen-antibody recognition, Lee et al. proposed an electrochemical immunosensor for selectively quantification of *Aspergillus niger via* detecting the extracellular proteins (**Figure 3B**), which relied on the immobilization of the extracellular proteins and its interference of redox cycling in interdigitated electrodes [52]. The utilization of secretion promoter at the sampling stage realized a highly sensitive response by 200-fold improvement for the *Aspergillus niger* monitoring in their previous study [53], which might attribute to the specific antibody recognition rather than to the amplified oxidation oxygen reduction reaction.

More importantly, current diagnosing methods for *Aspergillus* infection are not appropriate in clinical POC testing. To solve this concern, Yu et al. developed a novel simple and rapid analytical technique for DNA amplification based on loop-mediated isothermal amplification (LAMP) [54]. The LAMP approach was employed to detect the target gene TR34, a biomarker for *Aspergillus fumigatus* infection in patients. Assisted by the primer design of LAMP, this protocol contributed to rapid and selective identification of TR34, as well as the high detection sensitivity with 10 genomic copies per reaction. It was demonstrated that the TR34-LAMP platform can be considered as a POC diagnosis strategy for the screening of clinical *Aspergillus fumigatus* infection.

4. Applications of biosensing strategies for Aspergillosis

4.1 Gliotoxin-related biomarker

Genus *Aspergillus* is a group of fungi that can induce a majority of *Aspergillus* infection from allergic reaction to invasive diseases. Invasive aspergillosis (IA), one of the most severe and devastating *Aspergillus* infections, is defined as a rapid, acute, and life-threatening invasive disease with mortality rate as high as 90%. Of the various *Aspergillus* (such as *Aspergillus fumigatus*, *Aspergillus flavus*, *Aspergillus niger*), *Aspergillus fumigatus* is the most common and predominate one that causes IA. Gliotoxin, considered as the most toxic metabolites occurred by *Aspergillus fumigatus*, poses great hazards to immunosuppressed individuals. Therefore, the development of simple, rapid, sensitive, and point-of-care strategies for gliotoxin control is of high significance and urgency for early diagnosis of IA.

In the direction, assisted by the immobilization-free and GO (graphene oxide)-SELEX technique, Gao et al. proposed the pioneer work for the isolation and selection of the specific aptamer toward gliotoxin (**Figure 4A**) [55]. After the eighth selection cycle, fortunately, the ssDNA was enriched, and the aptamer APT8 was obtained and sequenced with a dissociated constant (K_D) of 376 nM. Then, the APT8 was further truncated into a shorter sequence consisting of only 24 nucleotides according to the mfold structure prediction. More encouragingly, the truncated aptamer APT8T1 was confirmed to recognize the gliotoxin with higher specificity (K_D = 196 nM) and could be further designed to APT8T1M with 18-fold improvement of K_D value. Accordingly, to validate the feasibility and selectivity of the aptamer, a simple fluorescent aptasensor was established for the detection of gliotoxin based on base pairing between the aptamer and its complementary DNA. The specific recognition of aptamer against gliotoxin caused the release of the aptamer/gliotoxin complex from the microplate and the fluorescent signal enhancement. The fluorescent signal was observed to be in linear relationship with levels of target gliotoxin in the range of 0.1–100 nM. The LOD was estimated to be 0.05 nM, which is significantly lower than the previous instrument methods like HPLC-MS/MS. Moreover, the successful application of the fluorescent aptasensor in human serum and urine samples demonstrated that this developed aptasensing platform offered a promising value in gliotoxin control. Inspired by this pioneer finding, combining the unique superiorities of MXene (Ti_3C_2) and TDNs (tetrahedral DNA nanostructures), Wang et al. developed a novel electrochemical aptasensor for highly efficient detection of gliotoxin based on nanomaterial functionalization (**Figure 4B**) [56]. Ti_3C_2 nanosheets, exhibiting large surface area, were modified with TDNs *via* the coordination interaction. The prepared nanocomposites allowed outstanding conductivity and molecule recognition toward the target for signal amplification. After the binding events between the aptamer and target, the cDNA was released and bound to the aptamer-modified nanocomposites, leading to

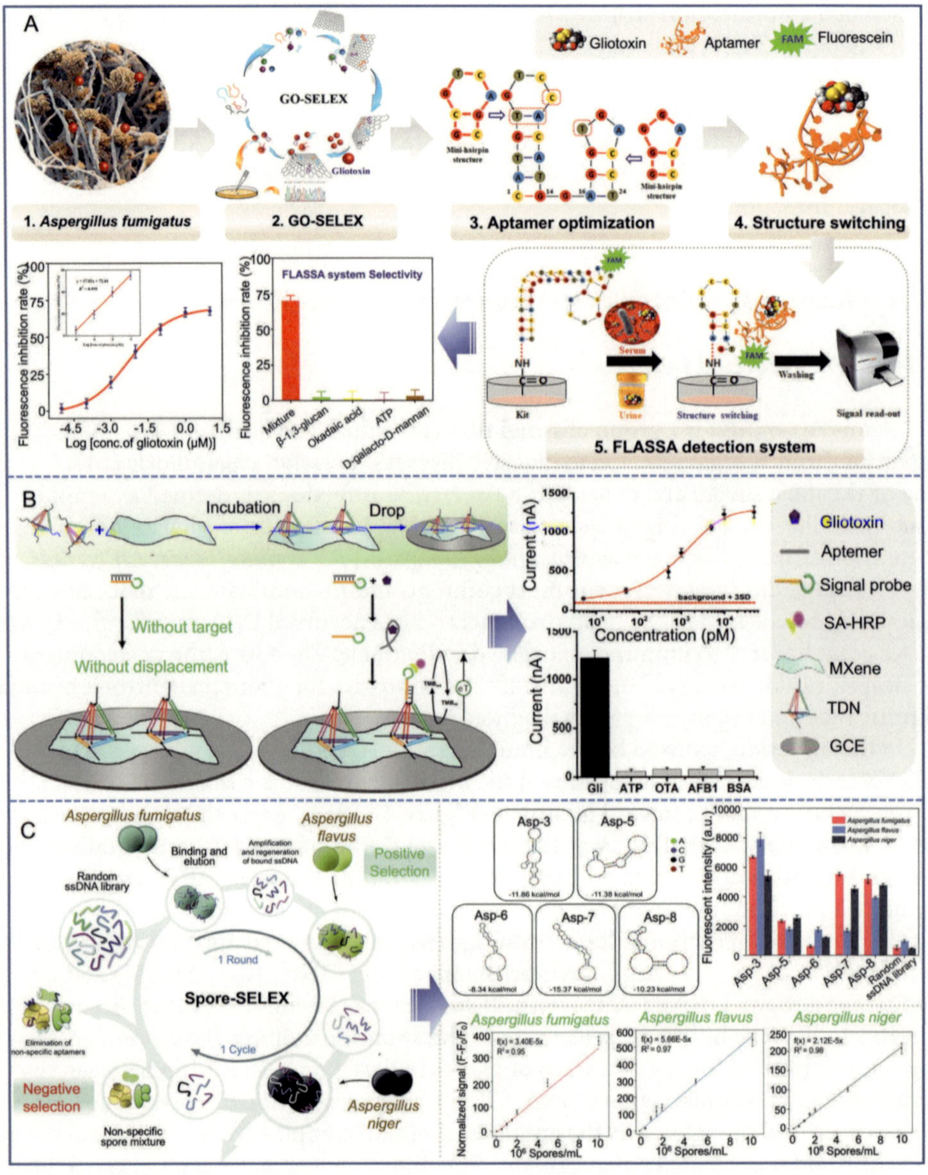

Figure 4.

(A) Working principle of the fluorescent aptasensor against gliotoxin detection for early diagnosis of invasive aspergillosis based on the specific aptamer selection. The illustration was recreated according to ref. [55]. Copyright 2018: American Chemical Society. (B) Schematic representation of electrochemical biosensor for label-free detection of gliotoxin incorporating of tetrahedral DNA nanostructures (TDNs) and MXene nanocomposites. The illustration was recreated according to ref. [56]. Copyright 2019: Elsevier. (C) Schematic diagram of isolation and characterization of the specific aptamer against spores of three representative Aspergillus *species. The illustration was recreated according to ref. [57]. Copyright 2021: Royal Society of Chemistry.*

the streptavidin-decorated HRP (horseradish peroxidase) catalysis and current signal generation. As a consequence, a dynamic response was achieved between the electrochemical signal and concentrations of gliotoxin ranged from 5 pM to 10 nM with an LOD of 5 pM. Accordingly, the detection sensitivity of this electrochemical aptasensor was improved by an order of magnitude than that of the previous fluorescent detection. Notably, the practicality of this method was excellent for gliotoxin detection in human serum samples. It was demonstrated that, from this viewpoint, advanced aptasensor technology possessed a significant potential against gliotoxin for early diagnosis of IA.

On the other hand, considering the obvious superiorities of aptamer-based biosensors, Seo et al. isolated and identified the specific aptamers that can specifically recognize *Aspergillus* spore by using cell-SELEX technique for the first time [57]. In this effort, the aptamers against three *Aspergillus* spores including *Aspergillus fumigatus*, *Aspergillus flavus*, and *Aspergillus niger* were selected in the spore-SELEX process (**Figure 4C**). With 12 rounds of selection, the ssDNA was successfully enriched, and the specific aptamer Asp-3 was achieved and sequenced with dissociated constants (K_D) of 80.12, 35.17, and 101.19 nM versus *Aspergillus fumigatus*, *Aspergillus flavus*, and *Aspergillus niger*, respectively. It was worth noting that the above aptamer also exhibited strong affinity to 1,3 β-D-glucans (BDGs) with K_D of 79.76–103.7 nM, demonstrating the excellent affinities and recognition potential toward *Aspergillus* spore surface molecules. However, the representative aptasensors have not been further developed. Hence, how to select and optimize the aptamer for *Aspergillus* spore control is of vital significance in the future. On the other hand, the biosynthesis of gliotoxin is related to gliP gene, and gliP gene-encoded enzyme further affects the occurrence of gliotoxin. In this direction, gliP gene can be considered as a potential biomarker for *Aspergillus* control, and it eventually contributed to the early diagnosis of IA. Encouragingly, Bhatnagar et al. fabricated an electrochemical biosensor toward gliP gene detection for the first time [58]. The gliP gene was immobilized onto chitosan-stabilized Au NPs on gold electrode. DNA hybridization reaction induced the formation of double-stranded DNA and its interaction with toluidine blue. The toluidine blue was acted as the electrochemical indicator for signal output. Upon the optimal conditions, the novel biosensor exhibited a dynamic response of the target in the range of 1×10^{-14}–1×10^{-2} M with an LOD of 0.32×10^{-14}. The proposed biosensor is stable and selective for detection of glip-T, demonstrating that it is useful for *Aspergillus* analysis and clinical diagnosis.

4.2 Galactomannan-related biomarker

GM (Galactomannan), regarded as a popular biomarker *Aspergillus* infection, is a heat-stable polysaccharide consisting of a linear mannan core with side chains of galactofuran. GM is metabolized and released in the blood and bronchoalveolar fluid (BALF) and is occurred as soluble antigen mainly in the cell wall of the genera *Aspergillus*. Ab (antibody)-based immunoassays such as ELISA (enzyme-linked immunosorbent assay), LFIA (lateral flow immunoassay), and immunosensor are the most frequently used protocols for the detection of GM in early infection of IA. Raval et al. firstly proposed an ELISA method to capture and detect GM based on the conjugation of polyclonal antibody to Au NPs [59]. The developed Au NPs immunoassay possessed simple and accurate detection of GM with low LOD at picomolar level (**Figure 5A**). Based on the similar mechanism, Guo et al. introduced a sandwich chemiluminescence immunoassay toward GM detection incorporated with luminescent nanomaterials (**Figure 5B**) [60]. In this attempt, the donor consisted of photosensitizer and phthalocyanine, and the acceptor contained the chemiluminescent dye, allowing the luminescent signal output under laser irradiation. The donor and the acceptor formed immunocomplex due to the close proximity less than 200 nm. The developed chemiluminescence ELISA platform displayed a linear response of GM in

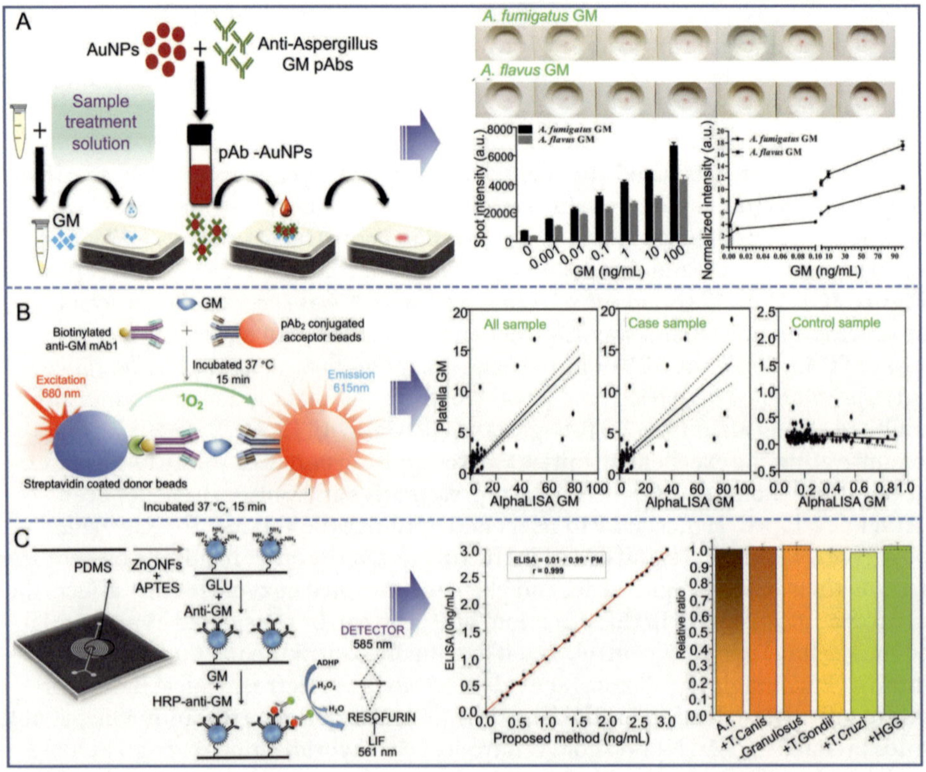

Figure 5.
(A) Schematic diagram of the ELISA-based immunoassay for the detection of galactomannan for early diagnosis of invasive aspergillosis based on the Au NPs conjugation. The illustration was recreated according to ref. [59]. Copyright 2019: Microbiology society. (B) Working principle of the chemiluminescence immunoassay toward galactomannan detection based on the donor and the acceptor. The illustration was recreated according to ref. [60]. Copyright 2022: Elsevier. (C) Schematic illustration of the fluorescence immunosensor to monitor galactomannan based on sandwich format and ZnO nanoflowers (ZnONFs). The illustration was recreated according to ref. [61]. Copyright 2020: Elsevier.

the range 0.05–100 ng/mL with an LOD of 0.032 ng/mL and was capable of highly efficient GM detection in serum and BALF. Besides, nanomaterials-based miniaturized biosensors possessed outstanding advantages in portable monitoring and high-throughput analytical manner [62–64]. Therefore, Piguillem et al. prepared the ZnO nanoflowers (ZnONFs), conjugated the nanomaterials to microfluidic channel, and employed the nanocomplex for antibody modification [61]. Correspondingly, a novel immunosensor was established for fluorescent detection of GM *via* the generation of fluorescent substance resorufin (**Figure 5C**). The immunoassay principle was designed by HRP-modified antibody for catalysis of 10-acetyl-3,7-dihydroxyphenoxacine oxidation to resorufin. Excitingly, the proposed immunosensor can realize a more sensitive and highly efficient detection of GM by 14-fold improvement in LOD over the commercial ELISA method.

5. Conclusions and outlook

Over the past decade, immunosensors and aptasensors have been well established for the detection and control of *Aspergillus* and Aspergillosis. Advanced nanomaterials-integrated approaches possess great potential to improve the performance for early diagnosis of IA. Of them, optical and electrochemical responses (including fluorescent, colorimetric, electrochemical, and electrochemiluminescence systems) are the two major signal transduction mechanisms. Besides, several researches have also focused on the principles toward binding events between the aptamer and target, catalytic transformation of target, truncation of the known aptamer, as well as the selection of new aptamers, and so on. These efforts indicated that advanced biosensors possessed excellent dynamic response and high sensitivity for AFB1 detection, as well as the feasibility and accuracy in *Aspergillus* detection and Aspergillosis diagnosis. Accordingly, biosensors techniques should have fascinating potential in industrial applications for food safety and risk assessment in the future.

Even though nanomaterials-based biosensors toward *Aspergillus* and Aspergillosis have witnessed the remarkable achievement, there are still several vital scientific limitations and challenges that required to overcome. (i) Advanced material-integrated biosensors exhibited excellent and improved analytical performance. For instance, as mentioned in this article, combining GO with AuNCs can significantly improve the fluorescent quenching efficiency compared to single GO. Novel luminescent materials such as UCNPs, PLNPs, and AIE probes displayed unique superiorities in terms of photostability, anti-interference, and even label-free detection. Therefore, the exploration of new materials and their synergistic effects are of great significance. (ii) Numerous reports are rarely aimed at developing the novel biosensors for their detection analysis rather than at the mechanism research. It is worth noting that the configuration change, site modification, and binding kinetics are of vital importance to better understand the principles and signal response. (iii) The selectivity assay is not appropriate in some case studies; the structural analogues should be taken into consideration for the interferences. (iv) It can be seen that the novel aptamer and aptasensors were relatively less developed, which is mainly attributed to the barriers in precise selection of aptamers. The stability and folding characteristics may be influenced by the special conditions like temperature, pH, ionic strength, and so on, thereby affecting the detection performance. Thus, how to select and identify high-quality aptamers that can undergo various reaction systems is a pursuing work. (v) Most established aptasensing protocols are limited in the laboratory conditions; in particular, at the present time, no commercial kits integrated by aptamers are available in the market. Enormous endeavors should be exerted to design and develop portable sensors or miniaturized devices in POC testing of *Aspergillus* and Aspergillosis.

Acknowledgements

We thank the University of Liège-Gembloux Agro-Bio Tech and more specifically the research platform AgricultureIsLife for the funding of the scientific stay in Belgium that made this paper possible.

Funding

This work was supported by the National Natural Science Foundation of China (No. 32001682, 21305158), the Special Fund for Agro-scientific Research in the Public Interest (201403071), and the Agricultural Science and Technology Research system of the PR China (ASTIP-IAS12).

Declaration of competing interest

The authors declare that they have no known competing financial interests or personal relationships that could have appeared to influence the work reported in this paper.

Author details

Xiaodong Guo[1,2]*, Mengke Zhang[3], Mengzhi Wang[1], Jiaqi Wang[4] and Marie-Laure Fauconnier[5]

1 Laboratory of Metabolic Manipulation of Herbivorous Animal Nutrition, College of Animal Science and Technology, Yangzhou University, Yangzhou, China

2 School of Agriculture and Biology, Shanghai Jiao Tong University, Shanghai, China

3 College of Food Science and Light Industry, Nanjing University of Technology, Nanjing, China

4 Laboratory of Quality and Safety Risk Assessment for Dairy Products of Ministry of Agriculture and Rural Affairs, Institute of Animal Science, Chinese Academy of Agricultural Sciences, Beijing, China

5 Chimie Générale et Organique, Gembloux Agro-Bio Tech, Université de Liège, Gembloux, Belgium

*Address all correspondence to: guoxiaodong233@163.com

References

[1] Farka Z, Jurik T, Kovář D, Trnkova L, Skládal P. Nanoparticle-based immunochemical biosensors and assays: Recent advances and challenges. Chemical Reviews. 2017;**117**(15):9973-10042. DOI: 10.1021/acs.chemrev.7b00037

[2] Wu J, Wang X, Wang Q, Lou Z, Li S, Zhu Y, et al. Nanomaterials with enzyme-like characteristics (nanozymes): Next-generation artificial enzymes (II). Chemical Society Reviews. 2019;**48**(4):1004-1076. DOI: 10.1039/C8CS00457A

[3] Chrouda A, Ayed D, Zinoubi K, Majdoub H, Jaffrezic-Renault N. Highly stable and ultra-sensitive amperometric aptasensor based on pectin stabilized gold nanoparticles on graphene oxide modified GCE for the detection of aflatoxin M1. Food Chemistry Advances. 2022;**1**:100068. DOI: 10.1016/j.focha.2022.100068

[4] Qi S, Duan N, Khan IM, Dong X, Zhang Y, Wu S, et al. Strategies to manipulate the performance of aptamers in SELEX, post-SELEX and microenvironment. Biotechnology Advances. 2022;**55**:107902. DOI: 10.1016/j.biotechadv.2021.107902

[5] Meirinho SG, Dias LG, Peres AM, Rodrigues LR. Voltammetric aptasensors for protein disease biomarkers detection: A review. Biotechnology Advances. 2016;**34**(5):941-953. DOI: 10.1016/j.biotechadv.2016.05.006

[6] Zhang M, Guo X. Gold/platinum bimetallic nanomaterials for immunoassay and immunosensing. Coordination Chemistry Reviews. 2022;**465**:214578. DOI: 10.1016/j.ccr.2022.214578

[7] Jiang G, Li Y, Liu J, Liu L, Pi F. Progress on aptamer-based SERS sensors for food safety and quality assessment: Methodology, current applications and future trends. Critical Reviews in Food Science and Nutrition. 2022:1-18. DOI: 10.1080/10408398.2022.2108370

[8] Qiao Q, Guo X, Wen F, Chen L, Xu Q, Zheng N, et al. Aptamer-based fluorescence quenching approach for detection of aflatoxin M1 in milk. Frontiers in Chemistry. 2021;**9**:653869. DOI: 10.3389/fchem.2021.653869

[9] Guo X, Qiao Q, Zhang M, Fauconnier ML. Nuclease triggered "signal-on" and amplified fluorescent sensing of fumonisin B1 incorporating graphene oxide and specific aptamer. International Journal of Molecular Sciences. 2022;**23**:9024. DOI: 10.3390/ijms23169024

[10] Hou Y, Jia B, Sheng P, Liao X, Shi L, Fang L, et al. Aptasensors for mycotoxins in foods: Recent advances and future trends. Comprehensive Reviews in Food Science and Food Safety. 2022;**21**(2):2032-2073. DOI: 10.1111/1541-4337.12858

[11] Zhai W, You T, Ouyang X, Wang M. Recent progress in mycotoxins detection based on surface-enhanced Raman spectroscopy. Comprehensive Reviews in Food Science and Food Safety. 2021;**20**(2):1887-1909. DOI: 10.1111/1541-4337.12686

[12] Park H, Baek S, Sen A, Jung B, Shim J, Park YC, et al. Ultrasensitive and selective field-effect transistor-based biosensor created by rings of MoS_2 Nanopores. ACS Nano. 2021;**16**(2):1826-1835. DOI: 10.1021/acsnano.1c08255

[13] Wei M, Rao H, Niu Z, Xue X, Luo M, Zhang X, et al. Breaking the time and space limitation of point-of-care testing strategies: Photothermometric sensors based on different photothermal agents and materials. Coordination Chemistry Reviews. 2021;**447**:214149. DOI: 10.1016/j.ccr.2021.214149

[14] Liu Y, Li B, Zhang H, Liu Y, Xie P. Participation of fluorescence technology in the cross-disciplinary detection of microcystins. Coordination Chemistry Reviews. 2022;**457**:214416. DOI: 10.1016/j.ccr.2022.214416

[15] Li Y, Liu D, Meng S, Zhang J, Li L, You T. Regulation of Ru $(bpy)_3^{2+}$ electrochemiluminescence based on distance-dependent electron transfer of ferrocene for dual-signal readout detection of aflatoxin B1 with high sensitivity. Analytical Chemistry. 2022;**94**(2):1294-1301. DOI: 10.1021/acs.analchem.1c04501

[16] Zhou Y, Yin H, Zhao WW, Ai S. Electrochemical, electrochemiluminescent and photoelectrochemical bioanalysis of epigenetic modifiers: A comprehensive review. Coordination Chemistry Reviews. 2020;**424**:213519. DOI: 10.1016/j.ccr.2020.213519

[17] Hu D, Hu Y, Zhan T, Zheng Y, Ran P, Liu X, et al. Coenzyme A-aptamer-facilitated label-free electrochemical stripping strategy for sensitive detection of histone acetyltransferase activity. Biosensors & Bioelectronics. 2020;**150**:111934. DOI: 10.1016/j.bios.2019.111934

[18] Lian X, Scott-Thomas A, Lewis JG, Bhatia M, MacPherson SA, Zeng Y, et al. Monoclonal antibodies and invasive Aspergillosis: Diagnostic and therapeutic perspectives. International Journal of Molecular Sciences. 2022;**23**:5563. DOI: 10.3390/ijms23105563

[19] Johnson G, Ferrini A, Dolan SK, Nolan T, Agrawal S, Doyle S, et al. Biomarkers for invasive aspergillosis: The challenges continue. Biomarkers in Medicine. 2014;**8**(3):429-451. DOI: 10.2217/bmm.13.129

[20] Suri CR, Boro R, Nangia Y, Gandhi S, Sharma P, Wangoo N, et al. Immunoanalytical techniques for analyzing pesticides in the environment. TrAC Trends in Analytical Chemistry. 2009;**28**(1):29-39. DOI: 10.1016/j.trac.2008.09.017

[21] Venkataramana M, Rashmi R, Uppalapati SR, Chandranayaka S, Balakrishna K, Radhika M, et al. Development of sandwich dot-ELISA for specific detection of Ochratoxin a and its application on to contaminated cereal grains originating from India. Frontiers in Microbiology. 2015;**6**:511. DOI: 10.3389/fmicb.2015.00511

[22] Jafari M, Rezaei M, Kalantari H, Tabarzad M, Daraei B. Optimization of aflatoxin B1 aptasensing. Journal of Toxicology. 2017;**2017**:2461354. DOI: 10.1155/2017/2461354

[23] Mozaffari Nejad AS, Sabouri Ghannad M, Kamkar A. Determination of aflatoxin B1 levels in Iranian and Indian spices by ELISA method. Toxin Reviews. 2014;**33**(4):151-154. DOI: 10.3109/15569543.2014.942319

[24] Sharma P, Gandhi S, Chopra A, Sekar N, Suri CR. Fluoroimmunoassay based on suppression of fluorescence self-quenching for ultra-sensitive detection of herbicide diuron. Analytica Chimica Acta. 2010;**676**(1-2):87-92. DOI: 10.1016/j.aca.2010.07.042

[25] Tey JN, Gandhi S, Wijaya IPM, Palaniappan A, Wei J, Rodriguez I, et al. Direct detection of heroin metabolites using a competitive immunoassay based

on a carbon-nanotube liquid-gated field-effect transistor. Small. 2010;**6**(9):993-998. DOI: 10.1002/smll.200902139

[26] Shim WB, Kim MJ, Mun H, Kim MG. An aptamer-based dipstick assay for the rapid and simple detection of aflatoxin B1. Biosensors & Bioelectronics. 2014;**62**:288-294. DOI: 10.1016/j.bios.2014.06.059

[27] Shim WB, Mun H, Joung HA, Ofori JA, Chung DH, Kim MG. Chemiluminescence competitive aptamer assay for the detection of aflatoxin B1 in corn samples. Food Control. 2014;**36**(1):30-35. DOI: 10.1016/j.foodcont.2013.07.042

[28] Gandhi S, Banga I, Maurya PK, Eremin SA. A gold nanoparticle-single-chain fragment variable antibody as an immunoprobe for rapid detection of morphine by dipstick. RSC Advances. 2018;**8**(3):1511-1518. DOI: 10.1039/C7RA12810J

[29] Gandhi S, Caplash N, Sharma P, Suri CR. Strip-based immunochromatographic assay using specific egg yolk antibodies for rapid detection of morphine in urine samples. Biosensors & Bioelectronics. 2009;**25**(2):502-505. DOI: 10.1016/j.bios.2009.07.018

[30] Talan A, Mishra A, Eremin SA, Narang J, Kumar A, Gandhi S. Ultrasensitive electrochemical immuno-sensing platform based on gold nanoparticles triggering chlorpyrifos detection in fruits and vegetables. Biosensors & Bioelectronics. 2018;**105**:14-21. DOI: 10.1016/j.bios.2018.01.013

[31] Ni SJ, Zhuo ZJ, Pan YF, Yu YY, Li FF, Liu J, et al. Recent progress in aptamer discoveries and modifications for therapeutic applications. ACS Applied Materials & Interfaces. 2021;**13**:9500-9519. DOI: 10.1021/acsami.0c05750

[32] Zon G. Recent advances in aptamer applications for analytical biochemistry. Analytical Biochemistry. 2022;**644**:113894. DOI: 10.1016/j.ab.2020.113894

[33] Aswani Kumar YVV, Renuka RM, Achuth J, Venkataramana M, Ushakiranmayi M, Sudhakar P. Development of hybrid IgG-aptamer sandwich immunoassay platform for aflatoxin B1 detection and its evaluation onto various field samples. Frontiers in Pharmacology. 2018;**9**:271. DOI: 10.3389/fphar.2018.00271

[34] Mukherjee M, Manonmani HK, Bhatt P. Aptamer as capture agent in enzyme-linked apta-sorbent assay (ELASA) for ultrasensitive detection of aflatoxin B1. Toxicon. 2018;**156**:28-33. DOI: 10.1016/j.toxicon.2018.11.001

[35] Kollia E, Proestos C, Zoumpoulakis P, Markaki P. Inhibitory effect of Cynara cardunculus L. extract on aflatoxin B1 production by aspergillus parasiticus in sesame (Sesamum indicum L). International Journal of Food Properties. 2017;**20**(6):1270-1279. DOI: 10.1080/10942912.2016.1206928

[36] Sardiñas N, Vázquez C, Gil-Serna J, González-Jaén MT, Patiño B. Specific detection and quantification of aspergillus flavus and aspergillus parasiticus in wheat flour by SYBR® green quantitative PCR. International Journal of Food Microbiology. 2011;**145**(1):121-125. DOI: 10.1016/j.ijfoodmicro.2010.11.041

[37] Rodrigues P, Soares C, Kozakiewicz Z, Paterson R, Lima N, Venâncio A. Identification and characterization of aspergillus Flavus and aflatoxins. Communicating

Current Research and Educational Topics and Trends in Applied Microbiology. 2007;**2**:527-534

[38] Wang C, Yuan X, Liu X, Gao Q, Qi H, Zhang C. Signal-on Impedimetric electrochemical DNA sensor using Dithiothreitol modified gold nanoparticle tag for highly sensitive DNA detection. Analytica Chimica Acta. 2013;**799**:36-43. DOI: 10.1016/j.aca.2013.09.024

[39] Sedighi-Khavidak S, Mazloum-Ardakani M, Khorasgani MR, Emtiazi G, Hosseinzadeh L. Detection of aflD gene in contaminated pistachio with aspergillus flavus by DNA based electrochemical biosensor. International Journal of Food Properties. 2017;**20**(S1):S119-S130. DOI: 10.1080/10942912.2017.1291675

[40] Liang M, Cai X, Gao Y, Yan H, Fu J, Tang X, et al. A versatile nanozyme integrated colorimetric and photothermal lateral flow immunoassay for highly sensitive and reliable aspergillus flavus detection. Biosensors and Biolelectronics. 2022;**213**:114435. DOI: 10.1016/j.bios.2022.114435

[41] Lin H, Wang F, Lin J, Yang W, Kang W, Jiang H, et al. Detection of wheat toxigenic aspergillus flavus based on nano-composite colorimetric sensing technology. Food Chemistry. 2023;**405**:134803. DOI: 10.1016/j.foodchem.2022.134803

[42] Ma J, Guan Y, Xing F, Eltzov E, Wang Y, Li X, et al. Accurate and non-destructive monitoring of mold contamination in foodstuffs based on whole-cell biosensor array coupling with machine-learning prediction models. Journal of Hazardous Materials. 2023;**449**:131030. DOI: 10.1016/j.jhazmat.2023.131030

[43] Lin H, Jiang H, He P, Haruna SA, Chen Q, Xue Z, et al. Non-destructive detection of heavy metals in vegetable oil based on nano-chemoselective response dye combined with near-infrared spectroscopy. Sensors & Actuators, B: Chemical. 2021;**335**:129716. DOI: 10.1016/j.snb.2021.129716

[44] Tietel Z, Melamed S, Lebiush S, Voet H, Namdar D, Eltzov E, Tsror (Lahkim) L. Characterization of volatile signatures of Pectobacterium and Dickeya spp. As biomarkers for early detection and identification - a major tool in potato blackleg and tuber soft rot management. LWT. 2021;**144**:111236. DOI:10.1016/j.lwt.2021.111236

[45] Josselin L, De Clerck C, De Boevre M, Moretti A, Jijakli MH, Soyeurt H, Fauconnier ML. Volatile organic compounds emitted by aspergillus flavus strains producing or not aflatoxin B1. Toxins. 2021;**13**(10): 705-724. DOI:10.3390/toxins13100705

[46] Loulier J, Lefort F, Stocki M, Asztemborska M, Szmigielski R, Siwek K, et al. Detection of fungi and oomycetes by volatiles using E-nose and SPME-GC/MS platforms. Molecules. 2020;**25**(23):5749-5776. DOI: 10.3390/molecules25235749

[47] Liu CJ, Yu H, Zhang BC, Liu SL, Liu CG, Li F, et al. Engineering whole-cell microbial biosensors: Design principles and applications in monitoring and treatment of heavy metals and organic pollutants. Biotechnology Advances. 2022;**60**:108019. DOI: 10.1016/j.biotechadv.2022.108019

[48] Monaci L, Pilolli R, De Angelis E, Godula M, Visconti A. Multi-allergen detection in food by micro high-performance liquid chromatography coupled to a dual cell linear ion trap mass spectrometry. Journal of Chromatography. A. 2014;**1358**:136-144. DOI: 10.1016/j.chroma.2014.06.092

[49] Koeberl M, Clarke D, Lopata AL. Next generation of food allergen quantification using mass spectrometric systems. Journal of Proteome Research. 2014;**13**:3499. DOI: 10.1021/pr500247r

[50] Svobodova M, Mairal T, Nadal P, Bermudo MC, O'Sullivan CK. Ultrasensitive aptamer based detection of β-conglutin food allergen. Food Chemistry. 2014;**165**:419. DOI: 10.1016/j.foodchem.2014.05.128

[51] Jin JH, Kim J, Jeon T, Shin SK, Sohn JR, Yi H, et al. Real-time selective monitoring of allergenic aspergillus molds using pentameric antibody-immobilized single-walled carbon nanotube-field effect transistors. RSC Advances. 2015;**5**:15728. DOI: 10.1039/C4RA15815F

[52] Lee J, Kim S, Chung HY, Kang A, Kim S, Hwang H, et al. Electrochemical microgap immunosensors for selective detection of pathogenic aspergillus Niger. Journal of Hazardous Materials. 2021;**411**:125069. DOI: 10.1016/j.jhazmat.2021.125069

[53] Kwon J, Cho EM, Nandhakumar P, Yang SI, Yang H. Rapid and sensitive detection of aspergillus Niger using a single-mediator system combined with redox cycling. Analytical Chemistry. 2018;**90**:13491-11349. DOI: 10.1021/acs.analchem.8b03417

[54] Yu LS, Rodriguez-Manzano J, Malpartida-Cardenas K, Sewell T, Bader O, Armstrong-James D, et al. Rapid and sensitive detection of azole-resistant aspergillus fumigatus by tandem repeat loop-mediated isothermal amplification. The Journal of Molecular Diagnostics. 2019;**21**(2):286-295. DOI: 10.1016/j.jmoldx.2018.10.004

[55] Gao S, Zheng X, Tang Y, Cheng Y, Hu., X., Wu, J. Development of a fluorescently labeled Aptamer structure-switching assay for sensitive and rapid detection of Gliotoxin. Analytical Chemistry. 2019;**91**(2):1610-1618. DOI: 10.1021/acs.analchem.8b05094

[56] Wang H, Li H, Huang Y, Xiong M, Wang F, Li C. A label-free electrochemical biosensor for highly sensitive detection of gliotoxin based on DNA nanostructure/MXene nanocomplexes. Biosensors & Bioelectronics. 2019;**142**:111531. DOI: 10.1016/j.bios.2019.111531

[57] Seo JW, Kim JY, Kim DH, Oh JJ, Kim YJ, Kim GH. Selection and characterization of toxic aspergillus spore-specific DNA aptamer using spore-SELEX. RSC Advances. 2021;**11**:2608-2615. DOI: 10.1039/D0RA09571K

[58] Bhatnagar I, Mahato K, Ealla KKR, Asthana A, Chandra P. Chitosan stabilized gold nanoparticle mediated self-assembled gliP nanobiosensor for diagnosis of invasive Aspergillosis. International Journal of Biological Macromolecules. 2018;**110**:449-456. DOI: 10.1016/j.ijbiomac.2017.12.084

[59] Raval KM, Ghormade V, Rajamohvotananet PR, Choudhary H, Rudramurthy SM, Chakrabarti A, et al. Developmentofanano-goldimmunodiagnostic assay for rapid on-site detection of invasive aspergillosis. Journal of Medical Microbiology. 2019;**68**(9):1341-1352. DOI: 10.1099/jmm.0.001040

[60] Guo Y, Li J. Sandwich-type homogeneous chemiluminescence immunoassay based on nanoparticle toward detection of aspergillus galactomannan antigen. Talanta. 2022;**243**:123392. DOI: 10.1016/j.talanta.2022.123392

[61] Piguillem SV, Regiart M, Bertotti M, Raba J, Messina GA, Fernández-Baldoet MA. Microfluidic fluorescence

immunosensor using ZnONFs for invasive aspergillosis determination. Microchemical Journal. 2020;**159**:105371. DOI: 10.1016/j.microc.2020.105371

[62] Wang HS, Wang C, He YK, Xiao FN, Bao WJ, Xia XH, et al. Core–Shell Ag@SiO_2 nanoparticles concentrated on a micro/Nanofluidic device for surface Plasmon resonance-enhanced fluorescent detection of highly reactive oxygen species. Analytical Chemistry. 2014;**86**(6):3013-3019. DOI: 10.1021/ac4037075

[63] Seia MA, Pereira SV, Fontán CA, De Vito IE, Messina GA, Raba J. Laser-induced fluorescence integrated in a microfluidic immunosensor for quantification of human serum IgG antibodies to helicobacter pylori. Sensors & Actuators, B: Chemical. 2012;**168**:297-302. DOI: 10.1016/j.snb.2012.04.026

[64] Fernández-Baldo MA, Ortega FG, Pereira SV, Bertolino FA, Serrano MJ, Lorente JA, et al. Nanostructured platform integrated into a microfluidic immunosensor coupled to laser-induced fluorescence for the epithelial cancer biomarker determination. Microchemical Journal. 2016;**128**:18-25. DOI: 10.1016/j.microc.2016.03.012

Chapter 6

Metagenomic Next-Generation Sequencing (mNGS) for the Diagnosis of Pulmonary Aspergillosis

Hao Tang, Shujun Bao and Caiming Zhong

Abstract

The diagnosis of pulmonary aspergillosis is a critical step in initiating prompt treatment and improving patients' prognosis. Currently, microbiological analysis of pulmonary aspergillosis involves fungal smear and culture, serum (1,3)-β-D-glucan (G) or galactomannan (GM) tests, and polymerase chain reaction (PCR). However, these methods have limitations. Recent studies have demonstrated that polymorphisms in pentraxin3 (PTX3), a soluble pattern recognition receptor, are associated with increased susceptibility to invasive aspergillosis. mNGS, a new microbial diagnostic method, has emerged as a promising alternative. It has high sensitivity in identifying pulmonary aspergillosis and can accurately distinguish species. Additionally, it outperforms other methods in detecting mixed infections and instructing the adjustment of antimicrobial treatments. As a result, mNGS has the potential to be adopted as the gold standard for the diagnosis of pulmonary aspergillosis.

Keywords: mNGS, fungi, pulmonary aspergillosis, diagnosis, other methods

1. Introduction

Aspergillus belongs to Ascomycetes fungi, including *Aspergillus* fumigatus, *Aspergillus* flavus, *Aspergillus* niger, and *Aspergillus* terreus, which are known to cause a wide range of diseases, such as allergic reactions, airway or lung infections, and extrapulmonary spread, especially skin infections. Lungs are the most common sites of *Aspergillus* infection in human. Meanwhile, pulmonary aspergillosis are often difficult to be diagnosed and have a high mortality rate. The clinical spectrum of pulmonary aspergillosis can present in various forms, including chronic pulmonary aspergillosis (CPA), aspergilloma, allergic bronchopulmonary aspergillosis (ABPA), and invasive pulmonary aspergillosis (IPA). The histopathology of the lesion can be understood as a phenotype that reflects the interaction between decreased host defense mechanisms and increased fungal virulence. In the case of *Aspergillus* infection, tissue reactions to the fungus are diminished, but the reasons and degree of the decrease in defense mechanisms vary among cases. Pulmonary aspergillosis presents

Type	IPA	CPA	ABPA	Aspergilloma
Main patient population	Patients with immune deficiencies; individuals without obvious immune deficiencies can also occur.	Patients with mild immune suppression caused by pulmonary diseases.	Hypersensitive reaction to *Aspergillus* antigens, which is commonly observed in patients with long-term asthma or cystic fibrosis.	Patients with normal immune function but pre-existing lung cavities (such as tuberculous cavities, lung cysts, bronchiectasis, etc.).
Crucial diagnostic methods	Chest CT, direct microscopy, histopathology, culture, as well as serum and BALF GM test	Combination of chest CT and specific IgG antibody detection for *Aspergillus*	Specific IgE antibody detection for *Aspergillus*	Chest CT

Table 1.
The main patient population and crucial diagnostic methods of different types of pulmonary aspergillosis.

in various forms in the disease spectrum depending on the interaction between *Aspergillus* and the host, and there may be an overlap between different clinical phenotypes. Moreover, the key diagnostic methods for different types of pulmonary aspergillosis vary. The main patient population and crucial diagnostic methods of different types of pulmonary aspergillosis are shown in **Table 1**.

In recent years, several new rapid diagnostic methods for pulmonary aspergillosis have emerged, in addition to traditional pathological and cultural diagnostic methods. These include the detection of antigens, antibodies, novel molecular markers, and genes. Among them, the second-generation sequencing technology for genetic testing has gained attention since 2014 [1], which has high sensitivity and a short detection cycle [2]. It can detect various samples such as sputum, bronchoalveolar lavage fluid (BALF), blood, and cerebrospinal fluid, making it a promising microbial identification technology. Which means that mNGS can classify a wide range of pathogens in several infection sites, including the respiratory tract [3], blood [4], central nervous system [5], and focal sites [6]. It is especially useful in detecting special and rare pathogens, as well as analyzing drug-resistance genes and virulence factors related to pathogens. The popularization of mNGS has proven to be of great value in diagnosing pulmonary aspergillosis.

2. The diagnostic value of mNGS in pulmonary aspergillosis

2.1 Principles and methods of mNGS

In recent years, the clinical application of mNGS has become an important method for diagnosing pulmonary fungal infections, which is of significant clinical value. Unlike traditional methods, mNGS technology does not require the culture of clinical samples. Instead, it extracts nucleic acid from all microorganisms in samples, constructs a standard sequencing library, and conducts high-throughput sequencing. Finally, the bioinformatics analysis confirms genome sequence comparison of each species, allowing for identification of microorganisms and calculation of various

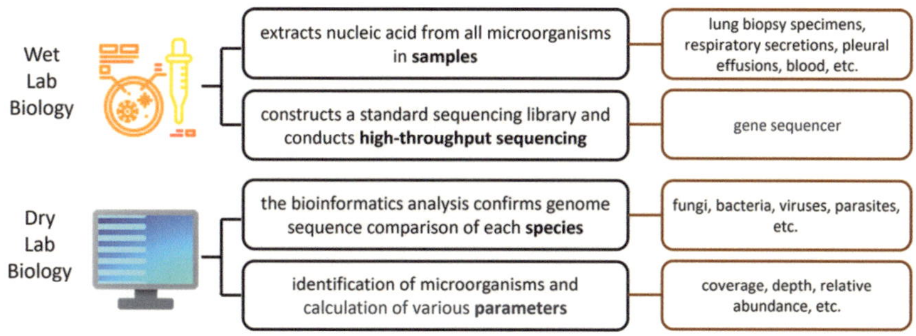

Figure 1.
Detection workflow of mNGS.

parameters. Please refer to **Figure 1**. mNGS has been utilized to identify the etiology of various clinical samples, significantly improving the diagnostic rate of pulmonary fungal infections. Several studies have demonstrated that mNGS is more sensitive than traditional tests, such as pathology, culture, serological tests, and PCR in fungal detection. Clinical studies have further confirmed the efficacy of mNGS in diagnosing common fungi and its ability to improve the sensitivity of pathogenic diagnosis of pneumonia fungi, making it an effective supplement to traditional microbiological detection methods for fungal infections.

2.2 Sample source for mNGS

Several studies have employed mNGS to detect microorganisms in lung biopsy specimens, respiratory secretions, and pleural effusions. For instance, mNGS was performed on formalin-fixed paraffin-embedded (FFPE) lung tissue samples from patients with granulomatous lesions and unclear diagnoses. Results showed that mNGS had a detection rate of 87.8% for total fungi and mycobacteria, which was 68.3% higher than histopathology. Moreover, pathogenic bacteria could be identified at the species level [7]. Other studies utilized mNGS to detect pathogenic microorganisms in computer tomography (CT)-guided lung biopsies of patients with lung diseases. Results showed that the specificity and positive predictive value of mNGS for detecting fungi were 100%, which was higher than histopathological methods [8].

Furthermore, the sensitivity of mNGS in detecting pathogenic bacteria and fungi in respiratory tract samples (such as nasopharyngeal swabs, sputum, and BALF) from patients with pulmonary infection was higher than traditional methods, such as respiratory virus culture and PCR analysis [9]. Other studies analyzed the diagnosis of pulmonary invasive fungal infections (IFIs) using mNGS and found that the fungal species detected by mNGS in BALF were higher than those detected by standard culture methods [10].

Some studies also tested patients with pulmonary infection combined with pleural effusion for mNGS, and the positive rate was higher than that of the culture method. Candida and Pneumospora were the most common fungal infections detected [11].

Bronchoscopic lung biopsy and BALF samples were also analyzed to investigate the differences in the detection of mNGS. Results showed that mNGS had a wider range of pathogen detection than traditional tests, especially for pulmonary fungal infection. The proportion of fungi detected and identified by mNGS was significantly higher than that by conventional tests [12]. Moreover, the sensitivity of mNGS in diagnosing pulmonary fungal infection was significantly higher than conventional tests, such as pathology, GM test, and culture. However, there was no difference in sensitivity and specificity between lung biopsy-mNGS and BALF-mNGS [13]. In other studies, transbronchial lung biopsy (TBLB), BALF, and bronchoalveolar needle brush (BB) specimens were tested for mNGS to detect suspected pulmonary infectious diseases. Results showed that the sensitivity of mNGS in detecting fungi in the three specimens was higher than that of conventional culture, but there was no difference in the sensitivity of mNGS among different specimen types [14].

According to the above researches, we can conclude that compared to the sensitivity of mNGS in the lung biopsy specimens, there is no obvious difference in that in respiratory secretions. Therefore, BALF/BB can be alternative samples in the detection of mNGS.

2.3 Fungi that can be detected by mNGS

The incidence of pulmonary fungal infections in clinical settings is attributable to a diverse range of pathogens such as *Aspergillus*, Mucor, Pneumospora [15], Cryptococcus, among others. Conversely, Candida is a relatively infrequent cause of pulmonary infections.

Lung infections associated with other pathogenic fungi such as Histoplasma capsulatum, Talaromyces marneffei, and Exophiala dermatitidis are even less common. mNGS has shown significant value in identifying rare fungi. It has been reported that a 27-year-old Chinese male with chronic progressive lung disease was asymptomatic for over a year until the disease progressed to the epiglottis, causing progressive pharyngeal pain. Despite negative results from BALF and epiglottic tissue cultures, as well as epiglottic and pulmonary pathology, mNGS was able to detect Histoplasma capsulatum in both epiglottic organizations and BALF and determined the cause after itraconazole treatment was successful [16]. In another case, mNGS successfully identified Talaromyces marneffei infection in a non-human immunodeficiency virus (HIV) patient, which is the first reported case in North China [17]. In the third case, a 52-year-old man with cough, sputum, and hemoptysis showed multiple lesions on both sides according to chest CT. Although the pathogen could not be identified after three biopsies, the subsequent mNGS results and therapeutic response confirmed that the pathogenic pathogen was Exophiala dermatitidis, and the diagnosis was Exophiala dermatitidis pneumonia [18].

Additionally, pulmonary fungal infections were frequently combined with bacterial infections, and immune deficiency and the presence of pulmonary bulba were risk factors associated with fungal and bacterial co-infections [19]. Furthermore, mNGS has been found to be able to distinguish colonization and infection in certain fungi, the colonization group and infection group of Pneumocystis yersii were differentiated by BALF-mNGS, and the fungal load was significantly different between the two groups [20].

2.4 The clinical application of mNGS in pulmonary aspergillosis

There are limited clinical studies on the diagnosis of pulmonary aspergillosis using mNGS. Although the studies are mostly limited to case reports, and the number of cases included are small, the available studies have demonstrated the good diagnostic performance of mNGS.

mNGS plays a key role in the diagnosis of pulmonary aspergillosis, either alone or in combination with other diagnostic methods. See **Figure 2** for details.

Patients with *Aspergillus* that can be detected by mNGS tend to have an incompetent immune function. One study reported that five patients' blood samples were positive for *Aspergillus* using mNGS, the five patients had various underlying diseases, including myelodysplastic syndrome (two cases), acute myeloid leukemia (two cases), and kidney transplantation (one case), which suggests that mNGS has guiding value in the diagnosis of IPA in immunocompromised patients [21]. In other studies, mNGS was used to detect plasma for the diagnosis of Corona Virus Disease 2019 (COVID-19)-associated pulmonary aspergillosis, suggesting that mNGS has good diagnostic performance for COVID-19-associated pulmonary aspergillosis when detecting plasma, and is highly specific [22]. Additionally, researchers detected mNGS in BALF samples and described three cases of IPA (respectively with chronic obstructive pulmonary diseases (COPD) and asthma history, and one case without underlying disease). *Aspergillus* fumigatus gene was found in all mNGS in BALF, indicating the diagnostic value of mNGS in non-neutropenic IPA patients [23]. Further, a study was conducted to analyze the accuracy of using mNGS to diagnose IPA in patients with different immune status. The study found that the consistency rate of IPA diagnosis in immunocompromised patients was significantly higher (82.1%) than in non-immunocompromised patients (52.9%). Therefore, mNGS detection combined with pulmonary CT imaging can be used for IPA diagnosis in patients with immunocompromised function. However, the diagnosis of IPA based on positive mNGS results in non-immunocompromised patients should be approached with caution [24].

Additionally, severe pneumonia caused by any type of pathogenic bacteria may be accompanied by *Aspergillus* infection. Therefore, some studies have used mNGS for early secondary infection screening, and secondary *Aspergillus* infection was found in a case of severe pneumonia caused by Legionella infection, leading to early precise treatment [25]. Another retrospective study found that mNGS performed well in the detection of common pathogens in *Aspergillus* infection, the most common bacteria were Klebsiella pneumoniae and Acinetobacter baumannii. Furthermore, 91.7% of pulmonary aspergillosis patients changed antibacterial therapy based on mNGS

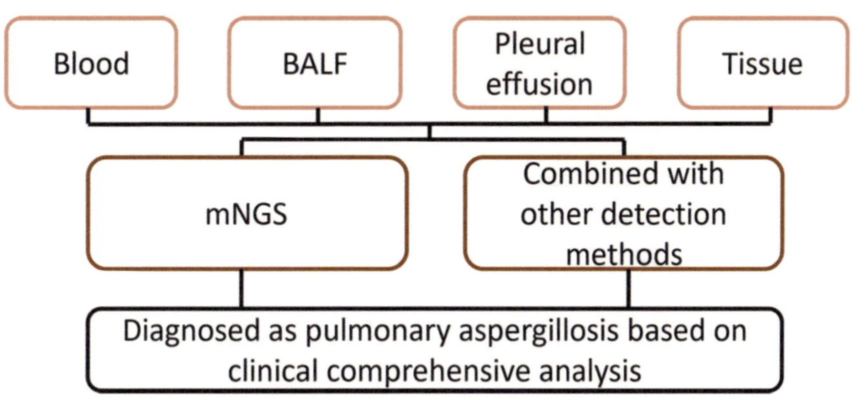

Figure 2.
The value of mNGS in the diagnosis of pulmonary aspergillosis.

Methods	Sensitivity (%)	Specificity (%)
Fungal smear	7.7	100
Culture	30.8	100
Serum(1,3)-β-D-glucan (G)	77.8–82.9	72.5–73.9
Galactomannan (GM)	57.7–77.8	90–92.9
PCR	64–86.7	84.2–99
mNGS	42.3–91.7	71.4–100

Table 2.
Comparing the sensitivity and specificity of different diagnostic methods of pulmonary aspergillosis.

results [26]. In summary, compared to other diagnostic methods, mNGS has higher sensitivity in the diagnosis of pulmonary aspergillosis. Please refer to **Table 2** for details.

mNGS can also be used as a supplement for routine microbial detection. A comparison of community-acquired pneumonia (CAP) patients diagnosed with IPA found that the sensitivity of GM to detect *Aspergillus* was the highest, followed by mNGS, culture, and smear. However, the specificity of mNGS, culture, and smear was 100%. The specificity of GM detection was 92.9% [27]. In patients with neutropenic, *Aspergillus* was detected by pathology and mNGS and diagnosed as IPA [28]. In patients with normal immune function, *Aspergillus* fumigatus was identified through endobronchial ultrasound-guided transbronchial needle aspiration (EBUS-TBNA) tissue with mNGS. Histological analysis of mediastinal biopsy and tissue fungal culture also indicated *Aspergillus* fumigatus infection, confirming the detection of mNGS [29]. In an infant patient initially suspected of having a lung tumor, mNGS identified *Aspergillus* fumigatus in BALF as the causative agent. The patient recovered quickly and was discharged after receiving appropriate antifungal therapy [30].

2.5 Precautions of mNGS report in the diagnosis of pulmonary aspergillosis

The interpretation of the mNGS report cannot disregard the value of a few or a dozen *Aspergillus* readings in the genus. This may occur when the thick cell wall of fungi makes nucleic acid extraction challenging. Additionally, after antifungal treatment, the number of fungi may decrease, resulting in low read lengths. In such cases, the sequenced samples may contain only free nucleic acid fragments that were lysed after the death of pathogenic bacteria. Moreover, the original pathogen load in the sample may have been low, leading to a low reading length. Consequently, some clinical studies fail to reflect the excellent diagnostic value of mNGS. In severe immune dysfunction patients suspected of pneumonia, the diagnostic accuracy of BALF's mNGS and conventional microbiological tests (CMTs) in detecting fungal infections was much lower. The low sensitivity to IPA was mainly responsible for this finding [31]. Furthermore, in rheumatic patients with suspected pneumonia and acute respiratory failure, the sensitivity of BALF sample culture combined with the GM test was superior to that of the mNGS test in detecting *Aspergillus* [32].

When low sequence numbers of microorganisms detected by mNGS are difficult to interpret clinically and cannot be verified by routine laboratory methods for clinical microorganisms, PCR detection may be used for confirmation. Combining mNGS with fungal PCR methods [33] can effectively detect fungi in clinical settings and reduce the rate of missed diagnoses. A small amount of *Aspergillus* detected by BALF-mNGS can be confirmed by BALF-GM and/or *Aspergillus*-specific IgG antibody. While histopathology is an objective criterion for detecting fungus, it cannot identify the species and should not be considered the ultimate criterion. At present, the best diagnostic method for pulmonary aspergillosis is a combination of lesion morphology with culture or sequencing.

If pulmonary fungal infection is detected only by mNGS, it could lead to significant confusion in clinical diagnosis. *Aspergillus* is commonly present in the environment and can colonize the respiratory tract of healthy individuals. A positive culture or molecular test does not necessarily indicate an infection, as *Aspergillus* nucleic acid sequences have been found in nearly 40% of samples from non-pulmonary fungal-infected individuals. Consequently, microbial sequences detected by mNGS may originate from pathogenic microorganisms, normal flora, transient colonization, sample contamination, or even incorrect database comparison.

3. Other new rapid diagnostic methods for pulmonary aspergillosis

The new rapid diagnostic methods for pulmonary aspergillosis include antigen and antibody detection, novel molecular markers, and gene detection.

Antigen detection methods mainly include the G test and GM test. The G test detects (1,3)-β-D-glucan, a specific component in fungal cell wall, as a pan-fungal detection biomarker. GM is a foreign antigen released by *Aspergillus* mycelium when invading host tissue, and it is a specific polysaccharide in cell wall of *Aspergillus*. The GM test is used for early *Aspergillus* infection diagnosis through an enzymatic-linked immunosorbent assay. The blood-GM test is relatively simple and easy to operate, but its sensitivity is low in non-neutropenic patients such as COPD [34]. Additionally, it is affected by antibacterial drugs [35], intravenous immunoglobulin [36], and multiple myeloma [37]. The BALF-GM test has better diagnostic value than serum-GM test does [34]. European and American guidelines recommend that non-neutropenic patients choose the BALF-GM test to improve the detection rate [38, 39]. However, its sensitivity and specificity of the BALF-GM test depend on the population included in the study, the defined cut-off point, the detection platform, and the time node, and whether *Aspergillus* covering drug therapy is performed. There is no universal diagnostic threshold.

Antibody detection includes IgG and IgE detection. IgG antibody detection is the most sensitive microbial test for the diagnosis of CPA [40]. The increase of specific IgE of *Aspergillus* and total IgE levels can confirm the diagnosis of ABPA.

Furthermore, the level of plasma PTX3 in IPA patients was significantly increased, which could be used as a novel molecular marker for diagnosis [41]. Additionally, PCR is a sensitive and rapid gene detection method, but has certain false positive and high negative predictive value. Studies have confirmed that BALF-PCR detection has a high diagnostic value in the diagnosis of invasive aspergillosis with immunocompromised function [42].

4. Conclusions

The mNGS technique has considerable diagnostic value in pulmonary aspergillosis, however, it should be closely integrated with clinical comprehensive judgment to achieve accurate diagnosis.

Acknowledgements

The authors would like to thank all patients for participating in this study. The authors also thank the BGI (Shanghai, China) for their helpful technical support. The research was sponsored by "Shuguang Program" supported by Shanghai Education Development Foundation and Shanghai Municipal Education Commission (20SG38), Shanghai Municipal Science and Technology Committee of Shanghai Outstanding Academic Leaders Plan (20XD1423300), and General Program of National Nature Science Foundation of China (No. 82070036).

Conflict of interest

The authors declare that the research was conducted in the absence of any commercial or financial relationships that could be construed as a potential conflict of interest.

Notes

All claims expressed in this article are solely those of the authors and do not necessarily represent those of their affiliated organizations, or those of the publisher, the editors and the reviewers. Any product that may be evaluated in this article, or claim that may be made by its manufacturer, is not guaranteed or endorsed by the publisher.

Nomenclature and abbreviations

mNGS	metagenomic next-generation sequencing
G	serum (1,3)-β-D-glucan
GM	galactomannan
PCR	polymerase chain reaction
PTX3	polymorphisms in pentraxin3
CPA	chronic pulmonary aspergillosis
ABPA	allergic bronchopulmonary aspergillosis
IPA	invasive pulmonary aspergillosis
BALF	bronchoalveolar lavage fluid
FFPE	formalin-fixed paraffin-embedded
CT	computer tomography
IFIs	invasive fungal infections
TBLB	transbronchial lung biopsy
BB	bronchoalveolar needle brush
HIV	human immunodeficiency virus
COVID-19	corona virus disease 2019
COPD	chronic obstructive pulmonary diseases
CAP	community-acquired pneumonia
EBUS-TBNA	endobronchial ultrasound-guided transbronchial needle aspiration
CMTs	conventional microbiological tests

Author details

Hao Tang*, Shujun Bao and Caiming Zhong
Department of Respiratory and Critical Care Medicine, Second Affiliated Hospital of Naval Medical University, Shanghai, China

*Address all correspondence to: tanghao_0921@126.com

References

[1] Wilson MR, Naccache SN, Samayoa E, Biagtan M, Bashir H, Yu G, et al. Actionable diagnosis of neuroleptospirosis by next-generation sequencing. The New England Journal of Medicine. 2014;**370**:2408-2417. DOI: 10.1056/NEJMoa1401268

[2] Ni PX, Ding X, Zhang YX, Yao X, Sun RX, Wang P, et al. Rapid detection and identification of infectious pathogens based on high-throughput sequencing. Chinese Medical Journal. 2015;**128**:877-883. DOI: 10.4103/0366-6999.154281

[3] Xie G, Zhao B, Wang X, Bao L, Xu Y, Ren X, et al. Exploring the clinical utility of metagenomic next-generation sequencing in the diagnosis of pulmonary infection. Infectious Diseases and Therapy. 2021;**10**:1419-1435. DOI: 10.1007/s40121-021-00476-w

[4] Jing C, Chen H, Liang Y, Zhong Y, Wang Q, Li L, et al. Clinical evaluation of an improved metagenomic next-generation sequencing test for the diagnosis of bloodstream infections. Clinical Chemistry. 2021;**67**:1133-1143. DOI: 10.1093/clinchem/hvab061

[5] Wilson MR, Sample HA, Zorn KC, Arevalo S, Yu G, Neuhaus J, et al. Clinical metagenomic sequencing for diagnosis of meningitis and encephalitis. The New England Journal of Medicine. 2019;**380**:2327-2340. DOI: 10.1056/NEJMoa1803396

[6] Thoendel MJ, Jeraldo PR, Greenwood-Quaintance KE, Yao JZ, Chia N, Hanssen AD, et al. Identification of prosthetic joint infection pathogens using a shotgun metagenomics approach. Clinical Infectious Diseases. 2018;**67**:1333-1338. DOI: 10.1093/cid/ciy303

[7] Sun W, Dong Z, Zhou Y, Xiong K, Liu H, Zhang Z, et al. Early identification of fungal and mycobacterium infections in pulmonary granulomas using metagenomic next-generation sequencing on formalin fixation and paraffin embedding tissue. Expert Review of Molecular Diagnostics. 2022;**22**:461-468. DOI: 10.1080/14737159.2022.2052046

[8] Li H, Gao H, Meng H, Wang Q, Li S, Chen H, et al. Detection of pulmonary infectious pathogens from lung biopsy tissues by metagenomic next-generation sequencing. Frontiers in Cellular and Infection Microbiology. 2018;**8**:205. DOI: 10.3389/fcimb.2018.00205

[9] Qian YY, Wang HY, Zhou Y, Zhang HC, Zhu YM, Zhou X, et al. Improving pulmonary infection diagnosis with metagenomic next generation sequencing. Frontiers in Cellular and Infection Microbiology. 2020;**10**:567615. DOI: 10.3389/fcimb.2020.567615

[10] Wang C, You Z, Fu J, Chen S, Bai D, Zhao H, et al. Application of metagenomic next-generation sequencing in the diagnosis of pulmonary invasive fungal disease. Frontiers in Cellular and Infection Microbiology. 2022;**12**:949505. DOI: 10.3389/fcimb.2022.949505

[11] Xu H, Hu X, Wang W, Chen H, Yu F, Zhang X, et al. Clinical application and evaluation of metagenomic next-generation sequencing in pulmonary infection with pleural effusion. Infection and Drug Resistance. 2022;**15**:2813-2824. DOI: 10.2147/idr.S365757

[12] Wang J, Han Y, Feng J. Metagenomic next-generation sequencing for mixed pulmonary infection diagnosis. BMC Pulmonary Medicine. 2019;**19**:252. DOI: 10.1186/s12890-019-1022-4

[13] Yang L, Song J, Wang Y, Feng J. Metagenomic next-generation sequencing for pulmonary fungal infection diagnosis: Lung biopsy versus bronchoalveolar lavage fluid. Infection and Drug Resistance. 2021;**14**:4333-4359. DOI: 10.2147/idr.S333818

[14] Wang Q, Wu B, Yang D, Yang C, Jin Z, Cao J, et al. Optimal specimen type for accurate diagnosis of infectious peripheral pulmonary lesions by mNGS. BMC Pulmonary Medicine. 2020;**20**:268. DOI: 10.1186/s12890-020-01298-1

[15] Liu N, Kan J, Yu N, Cao W, Cao J, Jiang E, et al. Application of metagenomic next-generation sequencing technology for difficult lung lesions in patients with haematological diseases. Translational Cancer Research. 2020;**9**:5245-5254. DOI: 10.21037/tcr-20-604

[16] Chen J, Li Y, Li Z, Chen G, Liu X, Ding L. Metagenomic next-generation sequencing identified Histoplasma capsulatum in the lung and epiglottis of a Chinese patient: A case report. International Journal of Infectious Diseases. 2020;**101**:33-37. DOI: 10.1016/j.ijid.2020.09.038

[17] Zhang J, Zhang D, Du J, Zhou Y, Cai Y, Sun R, et al. Rapid diagnosis of Talaromyces marneffei infection assisted by metagenomic next-generation sequencing in a HIV-negative patient. IDCases. 2021;**23**:e01055. DOI: 10.1016/j.idcr.2021.e01055

[18] Li Z, Tang J, Zhu J, Xie M, Huang S, Li S, et al. The convoluted process of diagnosing pulmonary mycosis caused by Exophiala dermatitidis: A case report. BMC Infectious Diseases. 2022;**22**:433. DOI: 10.1186/s12879-022-07399-y

[19] Zhao Z, Song J, Yang C, Yang L, Chen J, Li X, et al. Prevalence of fungal and bacterial co-infection in pulmonary fungal infections: A metagenomic next generation sequencing-based study. Frontiers in Cellular and Infection Microbiology. 2021;**11**:749905. DOI: 10.3389/fcimb.2021.749905

[20] Liu L, Yuan M, Shi Y, Su X. Clinical performance of BAL metagenomic next-generation sequence and serum (1,3)-β-D-Glucan for differential diagnosis of pneumocystis jirovecii pneumonia and pneumocystis jirovecii colonisation. Frontiers in Cellular and Infection Microbiology. 2021;**11**:784236. DOI: 10.3389/fcimb.2021.784236

[21] Ma X, Zhang S, Xing H, Li H, Chen J, Li H, et al. Invasive pulmonary aspergillosis diagnosis via peripheral blood metagenomic next-generation sequencing. Frontiers in Medicine. 2022;**9**:751617. DOI: 10.3389/fmed.2022.751617

[22] Hoenigl M, Egger M, Price J, Krause R, Prattes J, White PL. Metagenomic next-generation sequencing of plasma for diagnosis of COVID-19-associated pulmonary aspergillosis. Journal of Clinical Microbiology. 2023;**61**:e0185922. DOI: 10.1128/jcm.01859-22

[23] He BC, Liu LL, Chen BL, Zhang F, Su X. The application of next-generation sequencing in diagnosing invasive pulmonary aspergillosis: Three case reports. American Journal of Translational Research. 2019;**11**:2532-2539

[24] Zhan W, Liu Q, Yang C, Zhao Z, Yang L, Wang Y, et al. Evaluation of

metagenomic next-generation sequencing diagnosis for invasive pulmonary aspergillosis in immunocompromised and immunocompetent patients. Mycoses. 2023;**66**:331-337. DOI: 10.1111/myc.13557

[25] Yue R, Wu X, Li T, Chang L, Huang X, Pan L. Early detection of legionella pneumophila and aspergillus by mNGS in a critically Ill patient with legionella pneumonia after extracorporeal membrane oxygenation treatment: Case report and literature review. Frontiers in Medicine. 2021;**8**:686512. DOI: 10.3389/fmed.2021.686512

[26] Bao S, Song H, Chen Y, Zhong C, Tang H. Metagenomic next-generation sequencing for the diagnosis of pulmonary aspergillosis in non-neutropenic patients: A retrospective study. Frontiers in Cellular and Infection Microbiology. 2022;**12**:925982. DOI: 10.3389/fcimb.2022.925982

[27] Ao Z, Xu H, Li M, Liu H, Deng M, Liu Y. Clinical characteristics, diagnosis, outcomes and lung microbiome analysis of invasive pulmonary aspergillosis in the community-acquired pneumonia patients. BMJ Open Respiratory Research. 2023;**10**(1):e001358. DOI: 10.1136/bmjresp-2022-001358

[28] Hu Q, Li Y, Zhang Y, Sun S, Wang H, Jiang Z, et al. Case report: First report of T-cell large granular lymphocytic leukemia with NPL-DHX9 gene fusion successfully treated with cladribine: Clinical experience and literature review. Frontiers in Oncology. 2022;**12**:824393. DOI: 10.3389/fonc.2022.824393

[29] Deng W, Jiang Y, Qin J, Chen G, Lv Y, Lei Y, et al. Metagenomic next-generation sequencing assists in the diagnosis of mediastinal aspergillus fumigatus abscess in an immunocompetent patient: A case report and literature review. Infection and Drug Resistance. 2023;**16**:1865-1874. DOI: 10.2147/idr.S399484

[30] Yang A, Wang C, Chen P, Zheng G, Zhao Z, Liu J, et al. Diagnosis by metagenomic next-generation sequencing of invasive pulmonary aspergillosis in an infant with chronic granulomatous disease. Respiratory Medicine Case Reports. 2023;**41**:101792. DOI: 10.1016/j.rmcr.2022.101792

[31] Peng JM, Du B, Qin HY, Wang Q, Shi Y. Metagenomic next-generation sequencing for the diagnosis of suspected pneumonia in immunocompromised patients. The Journal of Infection. 2021;**82**:22-27. DOI: 10.1016/j.jinf.2021.01.029

[32] Shi Y, Peng JM, Qin HY, Du B. Metagenomic next-generation sequencing: A promising tool for diagnosis and treatment of suspected pneumonia in rheumatic patients with acute respiratory failure: Retrospective cohort study. Frontiers in Cellular and Infection Microbiology. 2022;**12**:941930. DOI: 10.3389/fcimb.2022.941930

[33] Ullmann AJ, Aguado JM, Arikan-Akdagli S, Denning DW, Groll AH, Lagrou K, et al. Diagnosis and management of Aspergillus diseases: Executive summary of the 2017 ESCMID-ECMM-ERS guideline. Clinical Microbiology and Infection. 2018;**24**(Suppl 1):e1-e38. DOI: 10.1016/j.cmi.2018.01.002

[34] Zhou W, Li H, Zhang Y, Huang M, He Q, Li P, et al. Diagnostic value of galactomannan antigen test in serum and bronchoalveolar lavage fluid samples from patients with nonneutropenic invasive pulmonary aspergillosis. Journal of Clinical Microbiology. 2017;**55**:2153-2161. DOI: 10.1128/jcm.00345-17

[35] Sulahian A, Touratier S, Ribaud P. False positive test for

aspergillus antigenemia related to concomitant administration of piperacillin and tazobactam. The New England Journal of Medicine. 2003;**349**:2366-2367. DOI: 10.1056/nejm200312113492424

[36] Liu WD, Lin SW, Shih MC, Su CL, Wang YW, Lin SC, et al. False-positive Aspergillus galactomannan immunoassays associated with intravenous human immunoglobulin administration. Clinical Microbiology and Infection. 2020;**26**:1555.e9-1555.e14. DOI: 10.1016/j.cmi.2020.02.002

[37] Ko JH, Peck KR, Lee JY, Cho SY, Ha YE, Kang CI, et al. Multiple myeloma as a major cause of false-positive galactomannan tests in adult patients with cancer. The Journal of Infection. 2016;**72**:233-239. DOI: 10.1016/j.jinf.2015.10.014

[38] Patterson TF, Thompson GR 3rd, Denning DW, Fishman JA, Hadley S, Herbrecht R, et al. Practice guidelines for the diagnosis and management of aspergillosis: 2016 update by the infectious diseases society of America. Clinical Infectious Diseases. 2016;**63**:e1-e60. DOI: 10.1093/cid/ciw326

[39] De Pauw B, Walsh TJ, Donnelly JP, Stevens DA, Edwards JE, Calandra T, et al. Revised definitions of invasive fungal disease from the European Organization for Research and Treatment of Cancer/Invasive Fungal Infections Cooperative Group and the National Institute of Allergy and Infectious Diseases Mycoses Study Group (EORTC/MSG) Consensus Group. Clinical Infectious Diseases. 2008;**46**:1813-1821. DOI: 10.1086/588660

[40] Page ID, Richardson MD, Denning DW. Comparison of six Aspergillus-specific IgG assays for the diagnosis of chronic pulmonary aspergillosis (CPA). The Journal of Infection. 2016;**72**:240-249. DOI: 10.1016/j.jinf.2015.11.003

[41] He Q, Li H, Rui Y, Liu L, He B, Shi Y, et al. Pentraxin 3 gene polymorphisms and pulmonary aspergillosis in chronic obstructive pulmonary disease patients. Clinical Infectious Diseases. 2018;**66**:261-267. DOI: 10.1093/cid/cix749

[42] Sun W, Wang K, Gao W, Su X, Qian Q, Lu X, et al. Evaluation of PCR on bronchoalveolar lavage fluid for diagnosis of invasive aspergillosis: A bivariate metaanalysis and systematic review. PLoS One. 2011;**6**:e28467. DOI: 10.1371/journal.pone.0028467

Section 4

Drug Development and Resistance

Chapter 7

Aspergilosis: Resistance and Future Impacts

Amanda Junior Jorge

Abstract

Fungal infections have been increasingly reported in routine, especially opportunistic ones such as aspergillosis, which represents a serious challenge for health professionals. The use of itraconazole, for a long time, was effective for a good clinical response, but factors associated with the advancement of medicine, length of stay, diagnostic errors, incorrect doses, and wrong choice of antifungal classes favored the appearance of resistance mechanisms. Thus, new research, together with the development of new molecules, is being carried out in order to reduce the advance of resistance, increasing patient survival.

Keywords: *Aspergillus*, therapeutics, ineffectiveness, antifungal, fungal infection

1. Introduction

Microorganisms of the genus *Aspergillus* are filamentous fungi, ubiquitous saprobes, and possess biological characteristics that allow their survival in temperature changes and extreme conditions, such as desert areas and polar regions. Its spores are much more resistant due to the development of thermotolerance, being able to survive up to 50°C. They have the ability to produce mycotoxins, such as glycotoxin, which have immunosuppressive activity, and the melanin pigmentation on the surface of the spores helps protect against UV rays [1].

Airway is the main form of infection (**Figure 1**), it occurs through inhalation of spores that are easily dispersed through the air, allowing their distribution over wide areas, such as open and closed environments, including hospitals. *Aspergillus* spores can be inhaled between 100 and 1000/day and reach the lung alveoli because of their reduced size (about 2–3 μm) [2]. However, they may not cause illness if inhaled by healthy individuals with competent immune systems. In turn, in immunocompromised individuals [3], the fungus lodges and causes lesions through the synthesis of enzymes (hemolysins, proteases, and peptidases) and toxins (fumagillin and gliotoxin) [4].

The proteases of some *Aspergillus* damage the protective barrier of the respiratory epithelium, inducing an inflammatory reaction to allow greater penetration of fungal antigens. They can also stimulate the release of pro-inflammatory cytokines (IL-8) and growth factors, causing bronchiectasis [4].

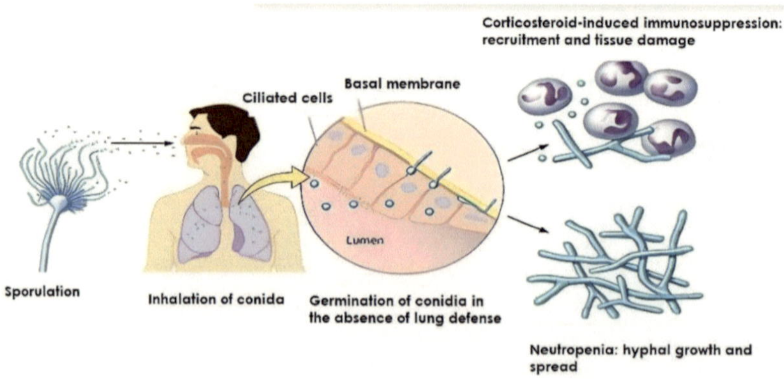

Figure 1.
Infectious cycle of Aspergillus *[2].*

One of the most severe forms is Invasive Pulmonary Aspergillosis, responsible for significant morbidity and mortality rates ranging from 80 to 100% in immunodeficient patients, with *A. fumigatus* being the main etiological agent. Some criteria for acquiring the disease are chronic obstructive pulmonary disease, asthma, hematologic malignancies such as acute myeloid leukemia (AML) or acute lymphoblastic leukemia (ALL), acquired immunodeficiency syndrome (AIDS), recipients of solid organ transplants and cirrhosis [5]. COPD is the main factor for the occurrence of API in 50% of cases, followed by solid organ transplants, because generally, in these patients, colonization with *Aspergillus* is 16.3 in every 1000 hospital admissions [6].

2. Discussion

As medical care has progressed, the number of patients at risk for invasive aspergillosis has increased. Bacterial treatments, intubation, and advances in medicine to prevent respiratory failure have also caused aspergillosis to evolve in a degree of pathogenicity [7]. In addition, the genus *Aspergillus* is much more frequent in solid organs; therefore, in transplanted individuals, caution must be exercised, and this is where we will most often find prophylactic therapy [8]. In general, to treat *Aspergillus*-related diseases are used polyenes, azoles, and echinocandins (**Table 1**) [9, 10].

The polyenes will bind to the ergosterol of the cytoplasmic membrane of the fungus, where they will change the permeability of the membrane, forming pores. These pores will allow the exit of proteins, carbohydrates, and cations, which will end up causing the death of the fungus, being, for this reason, a fungicidal drug [10].

The azoles represent an important class of chemicals for the management of fungal diseases in plants, animals, and humans and for the preservation of materials [11]. They are the first option for the treatment of aspergillosis [12] because of their azole ring that prevents the growth of fungi [13].

The mechanism of azoles is by blocking the cytochrome P-450-dependent enzyme lanosterol demethylase, affectin the ergosterol, a component of fungal plasma membranes [14]. Exposure to azoles in *A. fumigatus* decreases ergosterol levels, changing the membrane shape and structure, reducing the absorption of nutrients, chitin synthesis, and fungal growth [15, 16].

Antifungal class	Mechanism of action	Biological effect	Spectrum of action
Polyenes	Target ergosterol and extract sterols from fungal cell membranes	Fungicidal	Broad spectrum antifungal in treatment of invasive fungal infections; resistance is rare
Flucytosine	Inhibits DNA and RNA synthesis	Fungicidal against *Cryptococcus* spp.	Almost exclusively used for cryptococcal meningitis, but resistance is extremely common so never used in monotherapy
Azoles	Inhibit 14-α-lanosterol demethylase thereby inhibiting ergosterol synthesis	Mostly fungistatic	As a class they display broad spectrum against yeasts and filamentous fungi, although some species display intrinsic resistance to commonly used derivatives; secondary resistance can often develop during treatment
Echinocandins	Target 1,3-β-D-glucan synthase, thus preventing production of cell wall 1,3-β-D-glucan	Fungicidal against *Candida* spp., but fungistatic against *Aspergillus* spp.	First line of defense for candidiasis and used in aspergillosis when refractory to other treatments; resistance is emerging

Table 1.
The main classes of clinically-used antifungal agents for the treatment of invasive fungal infections [9].

Triazole is for long-term treatment, and it is the only anti-*Aspergillus* that can be administered orally [17]. For the treatment of noninvasive aspergillosis, itraconazole is used [18], although voriconazole is the first-line treatment for invasive aspergillosis [15, 16, 19].

Echinocandins is a new class of antifungal drugs, with the first example, caspofungin, entering clinical use a decade ago [20]. They are lipopeptide compounds that inhibit the enzyme [21, 22] beta-D-glucan synthase, which produces glucan, the main building block of fungal cell walls. It is indicated for patients with aspergillosis unresponsive to or intolerant of other treatments [20].

Regarding agriculture, plants are often attacked by various pathogenic fungi that cause a variety of diseases, such as leaf spot, blast, downy, etc. The stability of azoles is an important feature because even with small changes in their chemical composition, many azoles function for many days in agricultural habitats (soil and water). The azoles are effective against several fungal plant diseases [23]. Several foods contain azole residues; therefore, there is evidence that large amounts of antifungal residues, especially azoles, can remain in the environment [24].

Clinical resistance occurs when the maximum concentration of the drug is no longer efficient to eliminate the infection [25]. The emergence of the HIV pandemic, the increasing inappropriate use of drugs and illicit substances, transplant surgeries, abusive use of antifungals as prophylaxis for long periods in patients with immunosuppression, diagnostic failures, prolonged use in plantations, and factors inherent to the care units; intensive care such as mechanical ventilation, surgical interventions, total parenteral nutrition and prolonged treatment with antibiotics are some of the predisposing conditions for increased fungal infections by *Aspergillus* [26].

Increased resistance to azole therapy in patients with *Aspergillus* infections. *A. fumigatus*, which causes about 80% of invasive infections, has the highest resistance to azoles [27]. Infections with resistant *Aspergillus* strains cause ineffectiveness of azole antifungals, resulting in high mortality rates. In the United States, in the 1990s, strains resistant to *Aspergillus* had already been reported [28, 29], and since then, virtually all European nations have reported cases of azole resistance, including Germany, Ireland, Italy, Austria, Denmark, France, Sweden, Portugal, Spain, and Turkey [30, 31]. Studies began to be designed to analyze the resistance of *Aspergillus*, mainly *A. fumigatus*, which one carried out in the United Kingdom concluded that the resistance to azoles in *A. fumigatus* increased by 1,77% in 1998–2011 and 2015–2017 [32]. Another multicenter study in Taiwan found a high prevalence rate (4%) for azole-resistant *A. fumigatus* [33]. In addition, this study raised significant concerns about the use of azole antifungal drugs to treat invasive aspergillosis in the future.

Triazoles have been shown to have increased resistance due to an *Aspergillus* strain resistant to azoles after prolonged treatment in which mutations occur in the cyp51A codons [34]. Several mechanisms associated with cyp51A have been identified in patients on azole therapy [35]. Human-to-human transmission is therefore highly unlikely, and spread of resistance is very rare.

Soil is known to provide a natural habitat for a number of fungi that can be harmful including *Aspergillus*, *Coccidioides*, *Histoplasma*, and *Cryptococcus* [36]. Fungicides used repeatedly over a long period of time can create persistent selection pressure and lead to the development of resistant *Aspergillus* species. As a result, the environment contains *Aspergillus* species that are resistant to azoles. When susceptible individuals inhale these conidia, the *Aspergillus* species become resistant to the triazoles used for treatment. Several cases of triazole-resistant aspergillosis in humans and animals without prior triazole treatment have been reported worldwide [37, 38].

The mechanism of mutations induced in the environment in resistant *Aspergillus* comes from ergosterol (**Figure 2**), which is the component in a greater quantity of the cell membrane of fungi and is essential for the bioregulation of fluidity, asymmetry, and integrity of the cell membrane [10]. The cytochrome P450 enzyme, also called sterol-14α-demethylase, converts lanosterol to ergosterol. cyp51A is a gene that encodes the cytochrome P450 enzyme. The ergosterol biosynthetic pathway is the general target of azole antifungals. Triazoles prevent the cytochrome P450 enzyme from playing its role in converting lanosterol in the ergosterol biosynthetic pathway and cause ergosterol depletion and deleterious lanosterol accumulation [40]. Azole resistance is caused by mutations in the cyp51A gene that alter the cyp51A gene, protein structure, and reduce the enzymes' affinity for azole therapies [40].

Another factor associated with greater difficulty in achieving therapeutic success against aspergillosis is biofilm formation. Like biofilms on bacteria and yeast, *A. fumigatus* biofilms provide protection against antifungal and host immune defenses [41, 42].

Biofilms are formed by cells that adhere to abiotic and biotic surfaces and are surrounded by an extracellular matrix composed of polysaccharides. They act as a protective layer, aiding adhesion, surface integration, and cell propagation for subsequent invasion. This protection becomes less sensitive to treatments with antifungal drugs and attacks immune cells, making them more difficult to fight [43].

The *A. fumigatus* biofilm is highly resistant to all current classes of antifungal drugs, including azoles, echinocandins, and polyenes. The antifungal resistance associated with the *A. fumigatus* biofilm is thought to be a consequence of several interrelated factors, including elevated efflux pump activity, extracellular matrix production, and altered metabolic states [44].

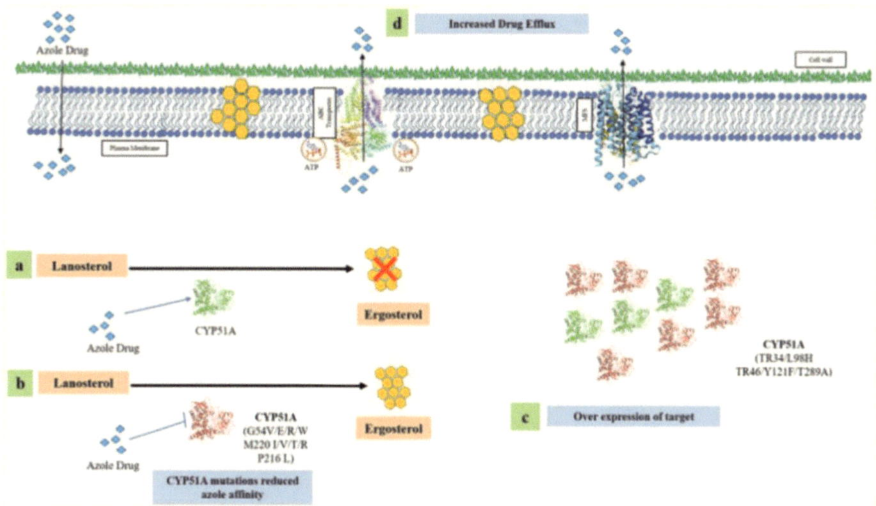

Figure 2.
Azole resistance mechanisms (a) wild-type fungi in the presence of azole drug unable to make ergosterol. (b) Mutations in the cyp51A region alter the structural modifications of the enzyme leading to reduce azole affinity. (c) Insertion of 34 and 46 bases pair in the promoter region along with point mutation in the cyp51A region causes overexpression of the gene. (d) Overexpression of efflux pump genes causes a reduced intracellular accumulation of azole drug [39].

When we compare the number of new antimicrobials with bacterial action, it can be seen that the development of antifungals faces challenges, as fungi present cellular similarities to the host, both are eukaryotic, and substances that will be toxic to the pathogen should not cause harm to the patient. Thus, the reduced number of antifungals currently used target structures belonging only to fungi [45].

Future therapeutic options aim to circumvent the existing limitations of current antifungal agents, and investigations have been carried out looking for targets different from those currently on the market, namely at the level of ergosterol, 1,3-β-D-glucan, and DNA. Such new approaches are favorable insofar as the toxicities and interactions may not be evidenced, as well as the resistances verified with other antifungal classes. New targets under development must be unique in addition to allowing cell viability [46].

A new option is the biosynthesis of glycosylphosphatidylinositol (GPI) AX001 (Amplyx Pharmaceuticals, San Diego, CA, USA), which consists of a new agent that, from the inhibition of inositol acyltransferase, mediated by the conserved fungal enzyme Gwt1, prevents the maturation of proteins linked to the GPI. Allowing agents to adhere to mucous membranes and epithelial surfaces, biofilm formation, and hyphal growth are crucial for colonization/infection. The main advantage of this molecule is its action only in fungal cells, having no activity in the acylation of human cells [45].

Its spectrum of action is quite broad, which allows it to act against *Candida*, *Aspergillus*, *Fusarium*, and *Scedosoporium*. It has no activity against *C. krusei* and *Mucorales* but has demonstrated *in vivo* efficacy against *Candida* species resistant to echinocandins and azoles. This molecule is undergoing a phase 1 trial of oral and intravenous formulations, seeking to assess its safety and tolerability. It has also been designated an orphan drug and Qualified Infectious Disease Product (QIDP) by the FDA [47].

The agent F901318, which is part of the orotomid class, has the ability to inhibit an oxidoreductase enzyme, dihydroorotate dehydrogenase, which interferes with the biosynthesis of pyrimidine. The inhibition by olorofim may affect the fungal cell wall and result in cell lysis. An *in vitro* study reported that exposure of A. fumigatus hyphae to olorofim (0.1 μg/mL for 24 h) led to significant reductions in 1,3-β-d-glucan at the hyphal tips and at the periphery of the mycelium [47].

Siderophores are iron chelators, so they manage to eliminate the available iron in various organisms such as plants, fungi, and bacteria. It is known that iron is essential for the viability of microorganisms, and if it is assimilated, it is no longer available for pathogens, constituting a good strategy. In this way, they manage to eliminate the iron present in the hosts that the agents may be infecting. VT-2397, formerly designated ASP2397, is isolated from Acremonium and permits aluminum chelation and was developed by Vical Pharmaceuticals (San Diego, CA, USA), demonstrating activity against azole-resistant *A. fumigatus* [48].

New molecules have also been developed based on the same targets: Azoles, with the elaboration of two molecules (VT-1161 and VT-1129) that are metalloenzymes and similar to azoles inhibit 14-α-demethylase, being directed to the treatment of infections by *Candida* and cryptococcal meningitis, respectively. What differentiates them from current azoles is the better selectivity, not binding to human CYP5, due to the fact that they have a tetrazole fraction in their structure, instead of triazole or imidazole, present in the agents available on the market [47].

Current echinocandins are only available in IV formulation. The SCY-078 molecule derived from enfumafungin has the advantage of being available in oral formulation, showing activity against several species of Candida and species resistant to fluconazole and in isolates with mutation at the level of the FKS1/FKS2 genes, which confer resistance to echinocandins. It demonstrated a spectrum of action comparable to commercial echinocandins, with emphasis on *C. glabrata* where it was eight times more effective. Recent studies have demonstrated activity against the new species *C. auris* [47, 48].

The CD101 molecule is also part of the echinocandins and has better solubility and less toxicity due to a modification at the choline level. It also demonstrates a much longer half-life and can be administered more widely – activity against *Candida* and *Aspergillus* species [47].

Regarding polyenes, the MAT2203 molecule (Matina BioPharma Holdings, Inc., USA) is a version of amphotericin B carried by nanoparticles that allows an oral formulation. In August 2015, it was approved by the FDA for the treatment of invasive candidiasis and aspergillosis. In phase 1 trial, it demonstrated a positive safety and tolerability profile [47].

3. Conclusion

Therefore, it was concluded that drugs for the treatment of fungal infections have undergone great advances in relation to pharmacodynamic properties, pharmacokinetics, spectrum of action, toxicity, and side effects. However, some factors contributed to an increase in the resistance profile and reduced therapeutic efficacy, and that is why researchers are developing new molecules based on the same targets as the available antifungal agents or new ones to circumvent clinical resistance. It is important to point out that the destination for antifungal effectiveness will not depend only on the synthesis of new drugs; it is also necessary to improve diagnoses, consider the benefits and harms of immunosuppressive therapies, and especially the choice of appropriate antifungal.

DOI: http://dx.doi.org/10.5772/intechopen.112755

Author details

Amanda Junior Jorge
Rua José Augusto Tourinho Dantas, Praia do Flamengo, Salvador, BA, Brazil

*Address all correspondence to: amandajjorge@gmail.com

IntechOpen

References

[1] Boral H, Metin B, Döğen A, Seyedmousavi S, Ilkit M. Overview of 8. Bibliografia 72 selected virulence attributes in aspergillus fumigatus, Candida albicans, Cryptococcus neoformans, Trichophyton rubrum, and Exophiala dermatitidis. Fungal Genetics and Biology. 2017;**111**:92-107. DOI: 10.1016/j.fgb.2017.10.008

[2] van de Veerdonk FL, Gresnigt MS, Romani L, Netea MG, Latgé J-P. Aspergillus fumigatus morphology and dynamic host interactions. Nature Reviews Microbiology. 2017;**15**(11): 661-674. DOI: 10.1038/nrmicro.2017.90

[3] Dagenais TRT, Keller NP. Pathogenesis of aspergillus fumigatus in invasive aspergillosis. Clinical Microbiology Reviews. 2009;**22**(3):447-465. DOI: 10.1128/CMR.00055-08

[4] Meireles MCA, Nascente PS. Micologia Veterinária. Ed. Universitária ed. Brazil: UFPel; 2009

[5] Koulenti D, Garnacho-Montero J, Blot S. Approach to invasive pulmonary aspergillosis in critically ill patients. Current Opinion in Infectious Diseases. 2014;**27**(2):174-183. DOI: 10.1097/ QCO.0000000000000043

[6] Guinea J, Torres-Narbona M, Gijón P, Muñoz P, Pozo F, Peláez T, et al. Pulmonary aspergillosis in patients with chronic obstructive pulmonary disease: Incidence, risk factors, and outcome. Clinical Microbiology and Infection. 2010;**16**(7):870-877. DOI: 10.1111/j.1469-0691.2009.03015.x

[7] Panackal AA. Combination antifungal therapy for invasive aspergillosis revisited. Medical Mycology Open Access. 2016;**2**(2):12. DOI: 10.21767/2471-8521.100012. Epub 2016 Apr 29. PMID: 27441304; PMCID: PMC4948747

[8] Rabagliati R. Actualización en el diagnóstico y manejo de aspergilosis invasora en pacientes adultos. Revista Chilena de Infectología. 2018;**35**(5):531-544. DOI: 10.4067/ s0716-10182018000500531

[9] Wall G, Lopes-Rabiot. Current antimycotics, new prospects, and future approaches to antifungal therapy. MPD Journal Antibiotics. 2020. DOI: 10.3390/ antibiotics9080445

[10] Kosmidis C, Denning DW. O espectro clínico da aspergilose pulmonar. Tórax. 2015;**70**(3):270-277. DOI: 10.1136/ thoraxjnl-2014-206291

[11] Fisher MC, Hawkins NJ, Sanglard D, Gurr SJ. Worldwide emergence of resistance to antifungal drugs challenges human health and food security. Science. 2018;**60**(6390):739-742

[12] Stevens DA, Moss RB, Kurup VP, et al. Participantes da Conferência de Consenso da Cystic Fibrosis Foundation. Aspergilose broncopulmonar alérgica na fibrose cística – estado da arte: Cystic Fibrosis Foundation Consensus Conference. Clinical Infectious Diseases. 2003;**37**(Suppl 3):S225-S264. DOI: 10.1086/376525

[13] Agentes antifúngicos Lass-Flörl C. Triazole em infecções fúngicas invasivas: uma revisão comparativa. Drogas. 2011;**71**(18):2405-2419. DOI: 10.2165/ 11596540-000000000-00000

[14] Snelders E, Karawajczyk A, Schaftenaar G, Verweij PE, Melchers WJ.

Perfil de resistência a azóis de alterações de aminoácidos em Aspergillus fumigatus CYP51A com base na modelagem de homologia de proteínas. Antimicrobial Agents and Chemotherapy. 2010;**54**(6):2425-2430. DOI: 10.1128/AAC.01599-09

[15] White TC, Marr KA, Bowden RA. Fatores clínicos, celulares e moleculares que contribuem para a resistência aos medicamentos antifúngicos. Clinical Microbiology Reviews. 1998;**11**(2):382-402. DOI: 10.1128/CMR.11.2.382

[16] Koltin Y, Hitchcock CA. A busca por novos agentes antifúngicos triazólicos. Current Opinion in Chemical Biology. 1997;**1**(2):176-182. DOI: 10.1016/S1367-5931(97)80007-5

[17] Maschmeyer G, Haas A, Cornely OA. Aspergilose invasiva: epidemiologia, diagnóstico e manejo em pacientes imunocomprometidos. Drogas. 2007;**67**(11):1567-1601. DOI: 10.2165/00003495-200767110-00004

[18] Howard SJ, Pasqualotto AC, Denning DW. Resistência a azóis na aspergilose broncopulmonar alérgica e bronquite por Aspergillus. Clinical Microbiology and Infection. 2010;**16**(6):683-688. DOI: 10.1111/j.1469-0691.2009.02911.x

[19] Walsh TJ, Anaissie EJ, Denning DW, et al. Sociedade de Doenças Infecciosas da América. Tratamento da aspergilose: diretrizes de prática clínica da Sociedade de Doenças Infecciosas da América. Clinical Infectious Diseases. 2008;**46**(3):327-360. DOI: 10.1086/525258

[20] Mukherjee PK, Sheehan D, Puzniak L, Schlamm H, Ghannoum MA. Equinocandinas: são todas iguais? Journal of Chemotherapy. 2011;**23**(6):319-325. DOI: 10.1179/joc.2011.23.6.319

[21] Patterson TF, Kirkpatrick WR, White M, et al. Aspergilose invasiva. Espectro da doença, práticas de tratamento e resultados. I3 Aspergillus Study Group. Medicine (Baltimore). 2000;**79**(4):250-260. DOI: 10.1097/00005792-200007000-00006

[22] Gupta AK, Sauder DN, Shear NH. Agentes antifúngicos: uma visão geral. Parte I. Journal of the American Academy of Dermatology. 1994;**30**(5 Pt 1):677-698. DOI: 10.1016/S0190-9622(08)81495-8

[23] Morton V, Staub T. A short history of fungicides. Online, APS*net* Features. 2008. DOI: 10.1094/APSnetFeature-2008-0308

[24] FRAC Code List ©*2022: Fungal control agents sorted by cross-resistance pattern and mode of action. Fungicide Resistance Action Committee. 2022

[25] Torres B. Mecanismos de resistencia, antifúngicos y nuevos antifúngicos rn desarrollo. Madrid: Faculdad de Farmacia, Universidad Complutense; 2018

[26] Botero MC, Puentes-Herrera M, Cortés JA. Formas lipídicas de anfotericina. Revista Chilena de Infectología. 2014;**31**(5):518-527. DOI: 10.4067/S0716-10182014000500002

[27] Messer SA, Jones RN, Fritsche TR. Vigilância internacional de Candida spp. e Aspergillus spp.: relatório do Programa de Vigilância Antimicrobiana SENTRY (2003). Journal of Clinical Microbiology. 2006;**44**(5):1782-1787. DOI: 10.1128/JCM.44.5.1782-1787.2006

[28] Gomez-Lopez A, Garcia-Effron G, Mellado E, Monzon A, Rodriguez-Tudela JL, Cuenca-Estrella M. Atividades in vitro de três agentes antifúngicos licenciados contra isolados

clínicos espanhóis de Aspergillus spp. Antimicrobial Agents and Chemotherapy. 2003;**47**(10):3085-3088. DOI: 10.1128/AAC.47.10.3085-3088.2003

[29] Lelièvre L, Groh M, Angebault C, Maherault AC, Didier E, Bougnoux ME. Aspergillus fumigatus resistente a azol: um problema emergente. Médecine et Maladies Infectieuses. 2013;**43**(4): 139-145. DOI: 10.1016/j.medmal.2013.02.010

[30] Heeres J, Backx LJ, Mostmans JH, Van Cutsem J, Antimicóticos imidazóis. Parte 4. Síntese e atividade antifúngica do cetoconazol, um novo potente agente antifúngico de amplo espectro oralmente ativo. Journal of Medicinal Chemistry. 1979;**22**(8):1003-1005. DOI: 10.1021/jm00194a023

[31] Mellado E, Garcia-Effron G, Alcázar-Fuoli L, Cuenca-Estrella M, Rodriguez-Tudela JL. Substituições na metionina 220 na 14α-sterol demetilase (Cyp51A) de Aspergillus fumigatus são responsáveis pela resistência in vitro aos antifúngicos azólicos. Antimicrobial Agents and Chemotherapy. 2004;**48**(7):2747-2750. DOI: 10.1128/AAC.48.7.2747-2750.2004

[32] Abdolrasouli A, Petrou MA, Park H, et al. Vigilância de Aspergillus fumigatus resistente a azóis em um serviço micológico de diagnóstico centralizado, Londres, Reino Unido, 1998-2017. Frontiers in Microbiology. 2018;**9**:2234. DOI: 10.3389/fmicb.2018.02234

[33] Wu CJ, Liu WL, Lai CC, et al. Estudo multicêntrico de isolados clínicos de Aspergillus fumigatus resistentes a azóis, Taiwan. Emerging Infectious Diseases. 2020;**26**(4):804-806. DOI: 10.3201/eid2604.190840

[34] Bueid A, Howard SJ, Moore CB, et al. Resistência antifúngica azólica em Aspergillus fumigatus: 2008 e 2009. The Journal of Antimicrobial Chemotherapy. 2010;**65**(10):2116-2118. DOI: 10.1093/jac/dkq279

[35] Howard SJ, Cerar D, Anderson MJ, et al. Frequência e evolução da resistência a azóis em Aspergillus fumigatus associada à falha terapêutica. Emerging Infectious Diseases. 2009;**15**(7):1068-1076. DOI: 10.3201/eid1507.090043

[36] Verweij PE, Chowdhary A, Melchers WJ, Meis JF. Resistência a azol em A. fumigatus: podemos manter o uso clínico de azóis antifúngicos ativos em fungos? Clinical Infectious Diseases. 2016;**62**(3):362-368. DOI: 10.1093/cid/civ885

[37] Meis JF, Chowdhary A, Rhodes JL, Fisher MC, Verweij PE. Implicações clínicas da resistência a azóis globalmente emergente em Aspergillus fumigatus. Philosophical Transactions of the Royal Society of London. Series B, Biological Sciences. 2016;**371**(1709):20150460. DOI: 10.1098/rstb.2015.0460

[38] Bunskoek PE, Seyedmousavi S, Gans SJ, et al. Tratamento bem-sucedido da aspergilose invasiva resistente a azóis em um golfinho nariz-de-garrafa com altas doses de posaconazol. Medical Mycology Case Reports. 2017;**16**:16-19. DOI: 10.1016/j.mmcr.2017.03.005

[39] Sen P, Vijay M, Singh S, et al. Understanding the environmental drivers of clinical azole resistance in Aspergillus species. Drug Target Insights. 2022;**16**:25-35. DOI: 10.33393/dti.2022.2476

[40] Snelders E, Camps SM, Karawajczyk A, et al. Fungicidas triazólicos podem induzir resistência cruzada a triazóis médicos em Aspergillus fumigatus. PLoS One. 2012;7(3):e31801. DOI: 10.1371/journal.pone.0031801

[41] Morelli KA, Kerkaert JD, Cramer RA. *Aspergillus fumigatus* biofilms: Rumo à compreensão de como o crescimento como uma rede multicelular aumenta a resistência antifúngica e a progressão da doença. PLOS Pathogens. 2021;**17**:e1009794

[42] Beauvais A et al. *Biofilme de Aspergillus* in vitro e *in vivo*. Medical Microbiology. 2015;**3**:4.08

[43] Dhingra S. Buckey, JC & Cramer, RA O oxigênio hiperbárico reduz a proliferação *de Aspergillus fumigatus* in vitro e influencia os resultados de doenças *in vivo*. Antimicrobial Agents and Chemotherapy. 2018;**62**:e01953-e01917

[44] Seidler MJ, Müller SS. FM *Aspergillus fumigatus* forma biofilmes com reduzida suscetibilidade a drogas antifúngicas em células epiteliais brônquicas. Antimicrobial Agents and Chemotherapy. 2008;**52**:4130-4136

[45] Roemer T, Krysan DJ. Antifungal drug development: Challenges, unmet clinical needs, and new approaches. Cold Spring Harbor Perspectives in Medicine. 2014;**4**:a019703

[46] Spitzer M, Robbins N, Wright GD. Combinatorial strategies for combating invasive fungal infections. Virulence. 2017;**8**(2):169-185. DOI: 10.1080/21505594.2016.1196300

[47] Gonzalez-Lara MF, Sifuentes-Osornio J, Ostrosky-Zeichner L. Drugs in clinical development for fungal infections. Drugs. 2017;**77**(14):1505-1518. DOI: 10.1007/s40265-017-0805-2

[48] Wiederhold NP. The antifungal arsenal: Alternative drugs and future targets. International Journal of Antimicrobial Agents. 2017;**51**(3):333-339. DOI: 10.1016/j.ijantimicag.2017.09.002

Section 5

Aflatoxins and Other *Aspergillus*-Related Metabolites

Chapter 8

Immunosensing of Aflatoxin B1 and Ochratoxin A on a Portable Device as Point-of-Care

Nur Azura Mohd Said, Noor Sheryna Jusoh, Norhafniza Awaludin, Mohammad Rejab Ismail, Noor Fadilah Mohd Bakri, Lily Suhaida Mohd Sojak and Faridah Salam

Abstract

Aflatoxin B1 (AFB1) and ochratoxin A (OTA) are potent mycotoxins produced by the fungal genus *Aspergillus*. Their occurrence in grain corn is alarming hence the need for rapid on-site detection. An immuno-based biosensor technique for detection of the aforementioned toxins is described here. Highly specific in-house polyclonal antibodies against AFB1 and OTA were employed as bioreceptors in a label-free electrochemical biosensor; immobilized on modified screen-printed carbon electrodes (SPCEs). The immuno-functionalized SPCEs were first characterized on a laboratory electrochemical workstation for proof-on-concept study using differential pulse voltammetry (DPV) electrochemical technique. An Android-based device is improvised as a portable electrochemical reader integrated with internet of thing (IoT) features which include cloud server and a dedicated website. Sensitivity achieved by the modified SPCEs on the portable device is superior compared to enzyme-linked immunosorbent assay (ELISA) method and lab-based electrochemical workstation. The miniaturized biosensor system has been successfully tested on cornfield for in-situ mycotoxins detection with simple sample extraction. Analysis performed on twenty samples were validated using chromatographic analysis. This biosensor-IoT system offers a potential application for real-time detection and the portable reader serves as an excellent tool for point-of-care in routine monitoring of harmful mycotoxins.

Keywords: aflatoxin B1, ochratoxin A, electrochemical biosensor, immunosensor, point-of-care, portable electrochemical device

1. Introduction

1.1 *Aspergillus* spp. and mycotoxins

Aspergillus is an oligotrophic fungus that can grow in an aerobic environment and is capable of producing a harmful secondary metabolite known as mycotoxin. In agriculture, mycotoxins contamination is one of the perturbing global food safety issues that need to be addressed. Its occurrence is widely spread infecting various important crops and commodities, [1] both during pre and postharvest conditions, and forages [2]; besides contaminating food products and animal feed. Among major crops that are susceptible to mycotoxin contaminations are peanuts, corn, rice, sorghum, chili, millets, and legumes [1]. Of recent, mycotoxins incidence in aquaculture also have been extensively reported, including fish feed [3, 4]. In food chain, storage environments such as poor aeration and high humidity are among factors that trigger and accelerate the growth of fungi, which consequently leads to mycotoxins production. Therefore, mycotoxins can exist in the field before harvest, postharvest, or during processing, storage, and feeding.

Among the most commonly observed mycotoxins that posed concerns to human health and livestock are aflatoxins, ochratoxin A, fumonisins, zearalenone, and deoxynivalenol. The harmful impacts of the mycotoxins have long been observed in both humans and animals [5]. Exposure to even low concentrations of mycotoxins in long term has been associated with liver diseases such as cancer, hepatitis, and jaundice, and to the extent of being carcinogenic. Aflatoxins and ochratoxins are principal toxins produced by *Aspergillus* spp. Human exposure to mycotoxins can happen either directly or indirectly. Directly, we are susceptible to mycotoxins when obliviously consuming grains, cereal-based products, or food contaminated with the toxigenic fungus. Indirectly, on the other hand, mycotoxins can enter human food chain *via* compromised animal feed. In terms of economic impact, farmers with contaminated crops or sick animals will lose their income and agribusiness will be greatly affected [6]. The prevalent issue of mycotoxins still remains the biggest challenge for animal feed producers and therefore, regular monitoring of mycotoxins is imperative.

1.1.1 Aflatoxins

The term aflatoxin is derived from the name of its main fungal producer that is *Aflavus*. Aflatoxins, also commonly known as a postharvest mold, are listed as potent mycotoxins. They are mainly produced by *Aspergillus flavus* and *A. parasiticus*. Other aflatoxin-producing species, albeit less frequently encountered, include *A. nomius*, *A. bombycis*, *A. pseudotamari,* and *A. ochraceoroseus*. Grains or food products contaminated with aflatoxins are not safe to be consumed. These toxins are resilient and stable against any thermal, physical, and chemical treatments along the feed chain production. In farm animals, aflatoxicosis can affect the liver and interrupt the digestive system. This in turn has a negative impact on livestock production with a reduction in body weight and feed conversion rate. Aflatoxins can contaminate a variety of livestock feeds and cause enormous economic losses, estimated at between US$52.1 and US$1.68 billion annually for the U.S. corn industry alone [7]. Even more unfortunate, aflatoxin in the animal's body system can be carried to eggs, meat, milk, and organ tissues. Indirect consumption of aflatoxins through food by humans can also cause toxin poisoning such as imperfect growth, weakened immune system, and liver damage, and have the potential to cause cancer and death.

There are more than 20 types of aflatoxins, but the four main types are aflatoxin B1 (AFB1), aflatoxin B2 (AFB2), aflatoxin G1 (AFG1), and aflatoxin G2 (AFG2). The terms B1, B2, G1, and G2 are addressed based on their fluorescence under ultraviolet light where B denotes blue and G denotes green color. Both AFB1 and AFB2 are produced by *A. flavus* while AFB1, AFB2, AFG1, and AFG2 are produced by *A. parasiticus* isolates. Meanwhile, aflatoxin M1 (AFM1) and aflatoxin M2 (AFM2) are hydroxylated metabolites of AFB1 and AFB2, respectively (**Figure 1**). Among all these types of aflatoxin, AFB1 has been classified as a class 1 human carcinogen by the International Agency for Research on Cancer [8]. The order of toxicity of aflatoxin is as follows: AFB1 > AFG1 > AFB2 > AFG2 [9]. Aflatoxins have been reportedly found in peanuts, cattle feed, liquid milk, cashew nuts [5], and also grain corn [10].

1.1.2 Ochratoxin

There are three classes of ochratoxins, namely ochratoxin A (OTA), ochratoxin B (OTB), and ochratoxin C (OTC) (**Figure 2**). They are produced by *Aspergillus* and

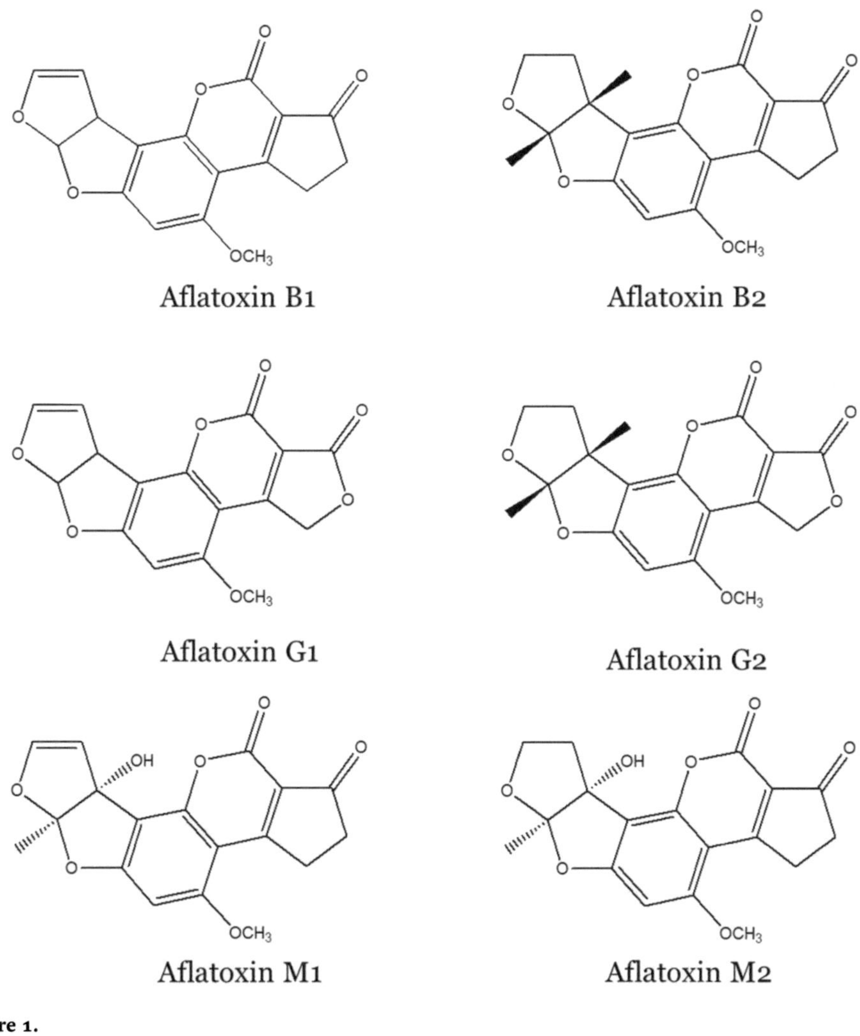
Aflatoxin B1 Aflatoxin B2
Aflatoxin G1 Aflatoxin G2
Aflatoxin M1 Aflatoxin M2

Figure 1.
Chemical structures of the main aflatoxins.

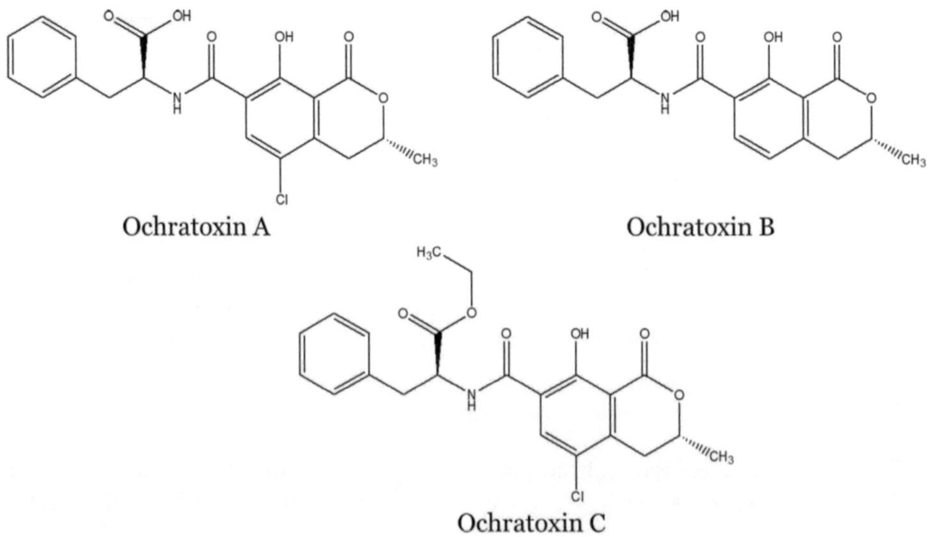

Figure 2.
Chemical structures of the ochratoxins.

Penicillium species, mainly *A. ochraceus*, *A. carbonarius*, *A. niger,* and *P. verrucosum*. Among these, OTA is considered the most abundant and harmful mycotoxin. OTA has been shown to be a potent nephrotoxic, hepatotoxic, and teratogenic compound. OTA has also been categorized as group 2B carcinogen for humans by the International Agency for Research on Cancer [8]. The intake of feed contaminated with OTA affects animal health and productivity and may result in the presence of OTA in the animal products. In acute cases, death may occur due to acute renal failure hence this toxin is regarded as nephrotoxic [11].

Ochratoxin commonly occurs in agricultural commodities, especially cereals this is corn and various foods such as cereal-based, wine, tea, coffee, cocoa, herbs, milk and milk products, poultry, fish, pork and eggs, fruits and vegetables, beans, dried products, infant foods as well as for poultry and other animal feeds. OTA is mostly found in the cereal grains such as maize, barley, oats, wheat, rye, etc. In a recent study, the abundance of ochratoxin A in grain corn is increasing in subregion of Asia in the second quarter of 2021 [10].

2. Mycotoxins and their detections

2.1 Current state detection for mycotoxins

Hitherto, instrumentation methods such as high-performance liquid chromatography (HPLC), liquid chromatography-mass spectrometry (LC-MS), and gas spectrometry (GC) are regarded as gold standard for food contaminants analysis, including mycotoxins. The chromatography system is usually coupled with detectors such as fluorescence, ultraviolet–visible, or mass spectrometry. Even though these techniques are sensitive, they require extensive sample extraction, trained personnel, and laborious. Antibody-based assay or immunoassays method for aflatoxins and ochratoxins detection include enzyme-linked immunosorbent assay (ELISA), chemiluminescent immunoassay (CLIA), fluorescence immunoassay (FIA), and lateral flow immunochromatographic assays (LFIA) [12]. Commercial ELISA kits also are widely available in the market for mycotoxins detection. However similar to chromatography analysis, ELISA method is still laboratory-bounded and requires a spectrophotometer to give accurate quantitative readings. Moreover, as ELISA is an optical-based technique, it is susceptible to light interference and hence is not suitable for on-site detection. LFIA strips also are being sought in the market as their analysis is straightforward, economical, and does not require skilled personnel. Nevertheless, the analysis is qualitative and only semiquantitative [13].

Besides immuno-based methods, spectrometric and spectroscopy techniques are also deemed sensitive for the detection of mycotoxin in food products. Near-infrared hyperspectral imaging has been successfully applied for the detection of fungal infection and OTA contamination in stored wheat and barley [14, 15]. Image processing techniques developed from machine learning is also gaining attention due to their nondestructive application for mycotoxins monitoring. Such has been applied for AFB1 in grains and feed commodities [16–19]. Machine learning methods, however, have limitations in terms of data volume, noises from the sensor, and calibration errors, besides being unable to ascertain types of mycotoxins produced.

Although tolerable limits for mycotoxins have been established, most contamination still exceeds maximum thresholds and hence continues to pose considerable risk to public health. As highlighted above, most of the reliable detections are lab-based and time-consuming. With this regard, when one requires immediate screening and on-site detection of mycotoxins, there is no readily available method. To address this, we have developed an immuno-based electrochemical biosensor for quantitative detection of *Aspergillus* spp. mycotoxins, namely AFB1 and OTA. Although the finding reported here is applied to grain corn (*Zea mays* L.), this developed biosensor platform can ultimately be applied to other food commodities.

2.2 Biosensor approach for mycotoxins detection

The first biosensor concept was introduced in 1962 based on an enzyme electrode for glucose sensing [20]. Ever since that, biosensor field has been receiving plethora of publications and vast applications in medicine [21–23], food safety [24, 25], agriculture [26–28], and environment [29]. By consensus, biosensor is defined as "*an analytical device, which exploits a biological detection or recognition system for a target molecule or macromolecule, in conjunction with a physicochemical transducer, which converts the biological recognition event into a useable output signal*" [30]. A biosensor system comprised of three essential components (**Figure 3**):

i. Detector—recognizes the biological element of interest (DNA, antibody, enzymes, cells, bacteria, etc.).

ii. Transducer—converts the biological element recognition to a readable signal.

iii. Output system—involves amplification and display of the signal.

Among the transducer mechanisms, electrochemical method is favorable in comparison to optical and mechanical. While optical biosensor is susceptible to light interference, electrochemical biosensor offers feasible application with simple sample extraction, rapid detection (within a few minutes), and point-of-care measurement.

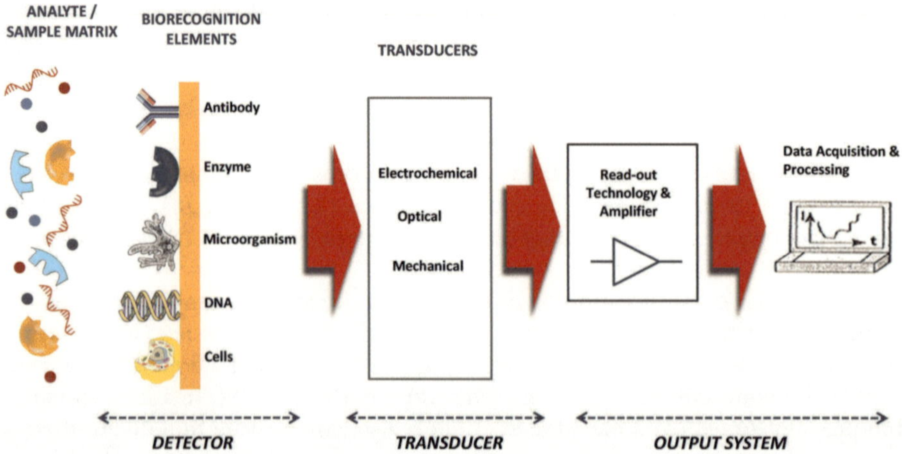

Figure 3.
Schematic diagram of biosensor comprising three components: Detector, transducer, and output system. Taken from ref. [31].

Main electrochemical sensing methodologies are voltammetry, potentiometry, impedimetric, conductometry, and amperometry. The recognition of biological elements (aptamer, enzyme, cell, antibody) with their targeted analytes by means of electrochemical measurement has been proven to provide reliability in numerous food contaminants detection, including mycotoxins in food and feed commodities [32–34]. Biosensors-based detection techniques have immense potential for mycotoxins detection in both research and industry as they are portable, simple, robust, sensitive, and cost-effective [35]. Of recent, aptamer, regarded as synthetic version of the antibody, also has been utilized in mycotoxin detection [36].

3. Strategies for advancing mycotoxins immunosensor

A miniaturized biosensor system with electrochemical tools and an integrated portable device for the detection of AFB1 and OTA mycotoxins is described in this study. The biosensor development employs in-house polyclonal antibody hence the term immunosensor. Research activities involved in mycotoxin immunosensor development are summarized in **Figure 4** as follows. In the first step, the antibody production is raised in rabbits as the animal host. After antibody purification and characterization, the antibody will be immobilized on screen-printed carbon electrodes (SPCEs), also addressed as sensor strips, for biosensing application. The immuno-functionalized SPCEs will be first optimized and tested on an electrochemical workstation for its proof-of-concept study. The strips are then attached to a handheld portable device integrated with I-of-Things (IoT) as point-of-care for rapid and real-time *in situ* analysis.

3.1 Components in biosensor integrated internet-of-things (IoT)

The integrated portable biosensor system for AFB1 and OTA detection with IoT system consisted of three main components (**Figure 5**): (i) an immuno-functionalized screen-printed carbon electrode (SPCE); (ii) a handheld electrochemical device with

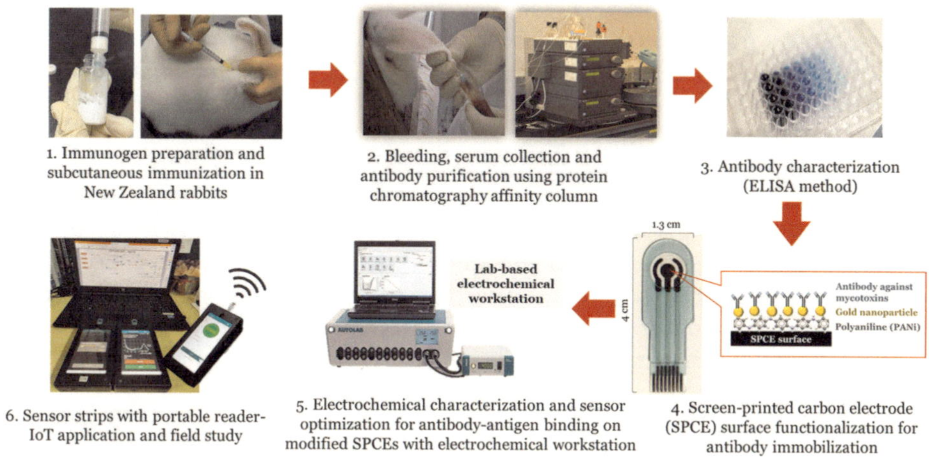

Figure 4.
Workflow for research activities involved in the development of immunosensor for mycotoxins detection.

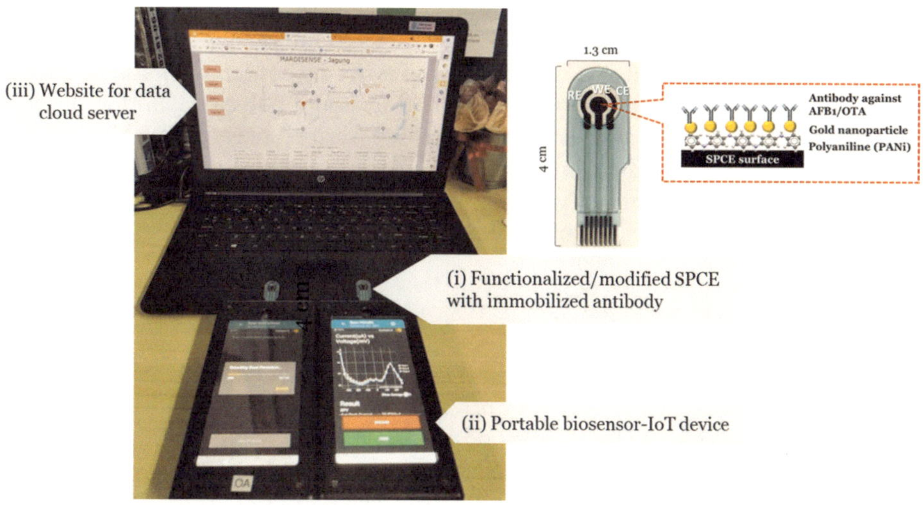

Figure 5.
Three main components of biosensor-IoT system. Inset: *A modified SPCE with polyaniline (PANi) and gold nanoparticle (AuNP) for antibody immobilization.*

electrochemical software; and (iii) a dedicated web server. The main part of the biosensor lies in the modified SPCEs. A plastic-based SPCE with single strip dimension of 1.3 cm (width) x 4 cm (length) is used in this study. The SPCEs have three carbon-based electrode channels: reference electrode (RE), working electrode (WE), and counter electrode (CE). WE is placed in the middle with a 4-mm circular well. The SPCEs are first modified with a conducting electroactive polymer (i.e. polyaniline, PANi) and nanogold (AuNP) network for antibody immobilization. The binding between the immobilized antibody and the targeted mycotoxin in samples will take place on the SPCE's WE surface.

Second component of the system is an improvised Android-based device with an electrochemical software and an electrode's scanner port as portable electrochemical reader. To perform the analysis, the SPCE is first inserted into the reader's port. The signal generated from antibody immobilized on the SPCE and its targeted mycotoxins is measured by the scanner on the electrochemical device *via* differential pulse voltammetry (DPV) technique. The quantified results in terms of toxin concentration (i.e. parts per billion, ppb) are displayed on the device's screen. This handheld device has built-in mobile app equipped with global positioning system (GPS) function, which allows sampling locations to be recorded. The third component, the web server, allows authorization and electrochemical settings of the portable device by registered personnel. As the portable reader is integrated with IoT, the server will collect all the field test data to allow users to perform further analysis such as location-mapping as well as trend-mapping. The system can also be set to prompt alerts to mobile phones if the test data exceeds certain preset limits to indicate immediate actions are required.

4. Development of electrochemical immunosensor for mycotoxins detection

4.1 Antibody as bioreceptor

Despite choices of aptamers and antibody as bioreceptors in mycotoxin biosensor, antibody still remains the most useful tool for identifying contaminants as the analysis can be performed on-site [6]. The production of in-house polyclonal antibody against AFB1 and OTA for mycotoxins immunosensor is described here. In biosensor development, polyclonal antibody is preferred over monoclonal antibody due to its multiple epitopes availability that can recognize its target analytes, hence signals generated are higher compared to those of monoclonal.

4.1.1 Production of polyclonal antibody against mycotoxins

Polyclonal antibodies against AFB1 and OTA were produced in-house at Biotechnology & Nanotechnology Research Centre, Malaysian Agricultural Research and Development Institute (MARDI). Antibody immunization against AFB1 and OTA was performed in four New Zealand rabbits (*Oryctolagus cuniculus*) with two replicates of rabbits for each toxin, denoted as A1 and A2 for AFB1, O1 and O2 for OTA. Animals protocol was reviewed and has been approved by the Animal Ethics Committee of MARDI (reference number 20190215/R/MAEC00045). For optimum antibody production and performance, immunization procedures were carried out adhering to the recommended guidelines [37].

Prior to immunization, blood was first taken and this preimmune batch is accounted as control study. Mycotoxins which have total molecular weight of >500 Da do not have immunogenicity property, hence as immunogen, they need to be coupled with a hapten in order to elicit immunity response in rabbits. Primary injection solution was prepared by homogenously mixing mycotoxins conjugated protein carrier with complete Freund's adjuvant (CFA) at 1:1 mixture. For subsequent immunogen, secondary injection solution was prepared by substituting CFA with incomplete Freund's Adjuvant (IFA). In total, eight immunization shots were performed and blood was collected until the fifth bleed within a duration of 5 months. Blood samples collected were purified with ammonium sulfate precipitation followed by nProtein A affinity column in order to obtain pure anti-AFB1 and anti-OTA IgG antibodies.

4.1.2 Characterizations of polyclonal antibody

Following antibody production and purification, the IgG needs to be further characterized to determine its performance, sensitivity, and selectivity. First, antibody titer is performed using indirect ELISA. The microtiters plates were coated with respective mycotoxins conjugated protein as the antigens. The performance of antibodies is evaluated by reducing the antibodies into several dilutions followed by optical measurement that is absorbance value. The purpose of titer is to determine which bleed of antibodies produced the best and most reliable result with the lowest dilution. These antibodies will then be used for further sensor development, thus, this step is exceptionally important. In general, a higher antibody titer indicates a better quality of fractionated antibody.

In both antibodies against AFB1 and OTA, preimmune antibodies that were used as control study showed no significant binding with the coated antigens (**Figures 6** and 7), indicating the successful production of the intended antibodies. From **Figure 6**, antibodies against AFB1 from third bleed onwards showed higher absorbance compared to the preimmune and the first two bleeds. The absorbance values recorded for both A1 and A2 rabbits exhibited similar range, indicating the reliability of the polyclonal antibody production in different hosts.

For antibodies against OTA, the absorbance of all bleeds (except preimmune) showed similar pattern until 10^{-3} dilution before absorbance reading gradually decreases upon further dilution (**Figure 7**). Antibodies from second bleed onwards

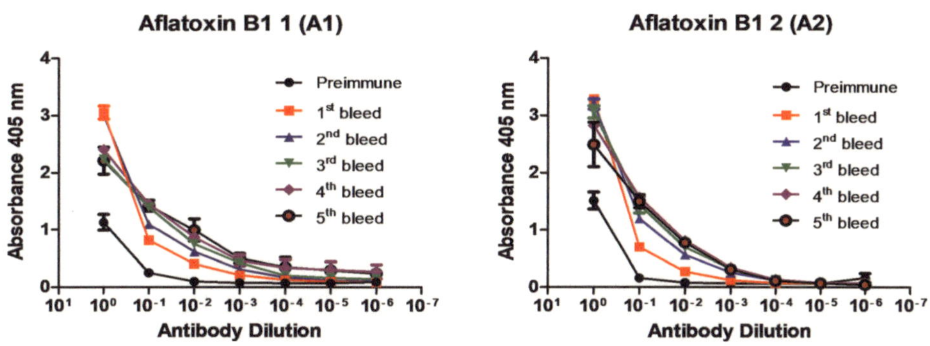

Figure 6.
Antibody titer for anti-AFB1 antibody produced from two different rabbits, A1 and A2.

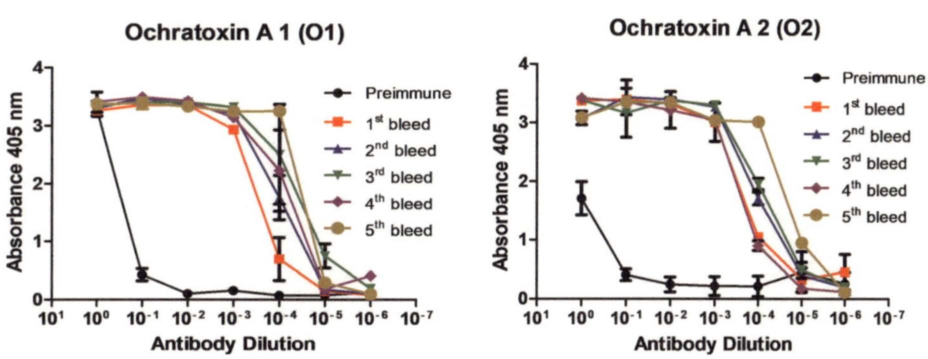

Figure 7.
Antibody titer for anti-OTA antibody produced from two different rabbits, O1 and O2.

showed the most prominent response compared to the first and preimmune bleed. Antibodies from third bleed exhibited the lowest concentration for antigen detection at 10^{-5} antibody dilution.

The amount of total protein in each antibody batch is then quantified by protein assay. Bicinchoninic acid or BCA protein assay is usually employed to determine the concentration of protein in the developed antibody. Bovine serum albumin (BSA) is used as the reference standard for antibody against mycotoxins assay. **Figure 8** displayed the concentration of antibody protein collected from different rabbits for anti-AFB1 and anti-OTA antibodies. The calculated concentration is based on BSA standard curve. Overall, the concentration was relatively low before introduction of mycotoxin into the rabbits' immune system. The protein concentration gradually increased as the rabbits built protective immune barrier against the mycotoxins after several injections. Some rabbits maintained the amount of antibodies in their bodies and some rabbits showed low tolerance of mycotoxins after four injections.

4.1.3 Cross-reaction studies

The purified IgGs are then subjected to cross-reaction study in assessing their specificity and sensitivity. The percentage of cross-reaction (CR) is obtained based on Eq. (1) while half-maximal inhibitory concentration (IC_{50}) values for linear response are calculated from Eq. (2). IC_{50} is one of the simplest and most accurate techniques to determine the performance of antibody toward specific target analyte in an assay. In this study, IC_{50} represents the 50% response of a series of mycotoxin concentrations (0–10 ppb) and corresponding electrochemical currents in a linear correlation, thus expressed in Eq. (2).

$$CR\ (\%) = \frac{IC50\ AFB1\ or\ OTA}{IC50\ toxins} * 100 \quad (1)$$

$$IC_{50} = \frac{(0.5 - c)}{m} \quad (2)$$

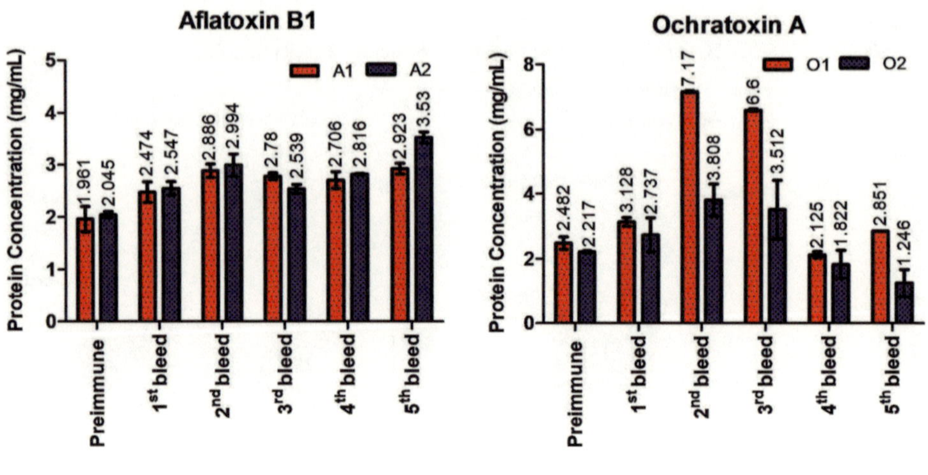

Figure 8.
Protein concentration found in antibodies against AFB1 (left) and OTA (right) for two rabbits.

The specificity of the anti-AFB1 polyclonal antibody produced was tested against other types of aflatoxins (i.e. AFB2, AFG1, and AFG2) and also other mycotoxins (ochratoxin A, ochratoxin B, fumonisin B1, fumonisin B2, and fumonisin B3). The anti-AFB1 antibody was found specific to AFB1, as indicated by the cross-reactivity studies shown in **Figure 9**. AFG1 showed cross-reaction with the antibody at 71% relative to AFB1 while AFB2 was at 65%. The cross-reaction study for AFG2 showed low cross-reactivity of less than 30%. The cross-reactivity order found was in the order of AFB1 > AFG1 > AFB2 > AFG2, corresponding with the order of toxicity reported [9, 38]. Meanwhile, when tested against other mycotoxins, anti-AFB1 antibody exhibited no cross-reaction with ochratoxins A and B, and fumonisins B1, B2, and B3. Overall, the in-house polyclonal antibody produced is highly sensitive toward AFB1, displaying high reactivity toward same group compounds and minimal cross-reactivity toward different group compounds. Similarly, anti-OTA antibody produced displayed nonsignificant cross-reactivity toward all groups of mycotoxins (**Figure 9**). The performance of OTA antibody in competitive assay was highly selective toward its respective target analyte.

4.2 Modified SPCEs for mycotoxins detection

Screen-printed electrodes (SPEs) fabricated by means of thick film deposition lend themselves well in biosensor construction. The miniaturization of three-cell electrodes integrated in one chip offers an economical, straightforward analysis and practicality that suit on-site analysis application [39]. The incorporation of nanomaterials and conducting polymers on the SPE's surface has successfully enhanced the sensitivity of the detection system [40, 41]. Here, the working electrode (WE) area on the SPCE was first modified by drop-casting a mixture of PANi and AuNP for antibody immobilization.

The combination of PANi and AuNP network was found to provide numerous binding sites for biomolecules and enhance current signals due to the Au nano-size and the properties of the conducting polymer, respectively. Antibody concentration of 0.1 mg/mL, which showed the most optimum electrochemical response toward the target analyte, was immobilized on the carbon WE surface (**Figure 10a**). The

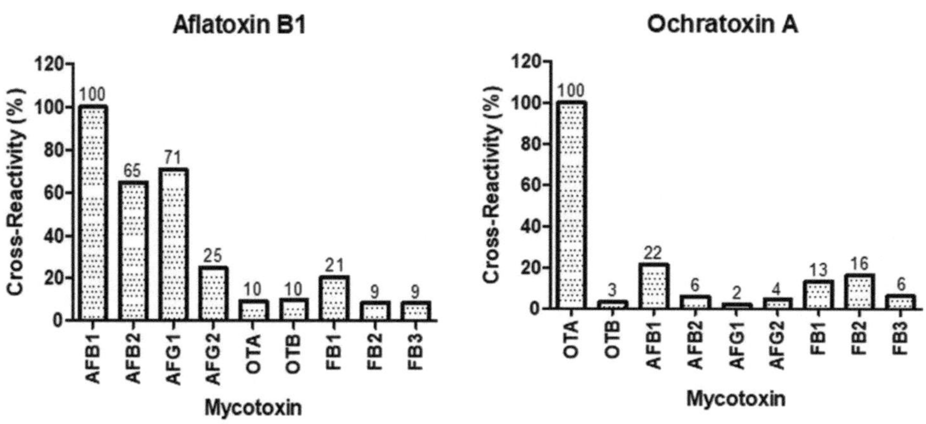

Figure 9.
Specificity study of anti-AFB1 antibody (left) and anti-OTA antibody (right) with their targeted mycotoxins and other mycotoxins groups. The percentage of cross-reaction was calculated based on IC_{50} readings.

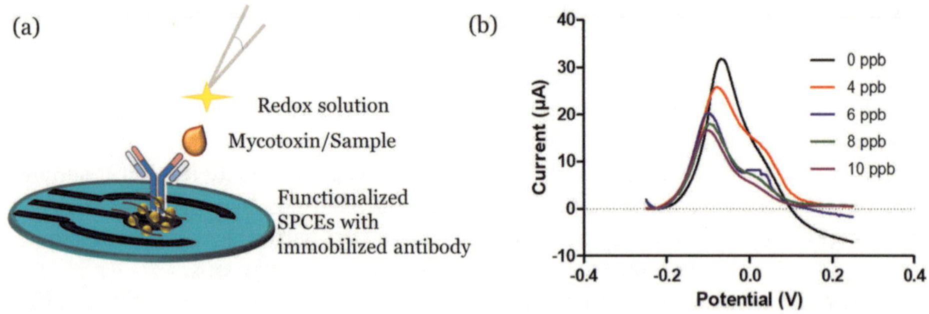

Figure 10.
(a) Working electrode (WE) of SPCE surface modification for antibody immobilization; and (b) differential pulse voltammetry (DPV measurement) for label-free approach for mycotoxin immunosensor.

electrochemical characterization from the SPCE modification has been detailed elsewhere [42]. The label-free electrochemical biosensor approach by means of DPV measurement (**Figure 10b**) was carried out by applying a redox solution of 5 mM ferricyanide/ferrocyanide in 0.1 M KCl on the SPCEs.

4.3 Sensor optimizations and standard curves development

Processed and nonprocessed grain corns can produce harmful mycotoxins along the production chain which can lead to serious health issues for livestock and human. In grain corn, *Aspergillus* species are the main producers of mycotoxins in maize, infecting both preharvest and during storage. Standard curves for AFB1 and OTA have been developed in grain corn as sample matrix model (**Figure 11**). The electrochemical measurement was carried out in undiluted and interference-free sample matrix on an electrochemical workstation. Peak current from voltammogram for each concentration was selected as the corresponding antibody-antigen response from electrochemical process and plotted on the graph. Overall, both AFB1 and OTA displayed a good linear correlation (R2) of 0.971 and 0.9685, respectively, in a broad working range of 0–10 ppb.

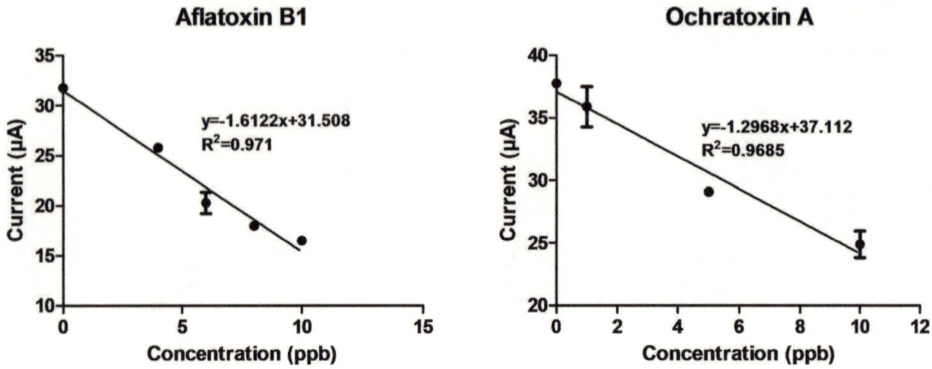

Figure 11.
Linear response from electrochemical workstation for a series of AFB1 (left) and OTA (right) concentrations with current reading in grain corn matrix.

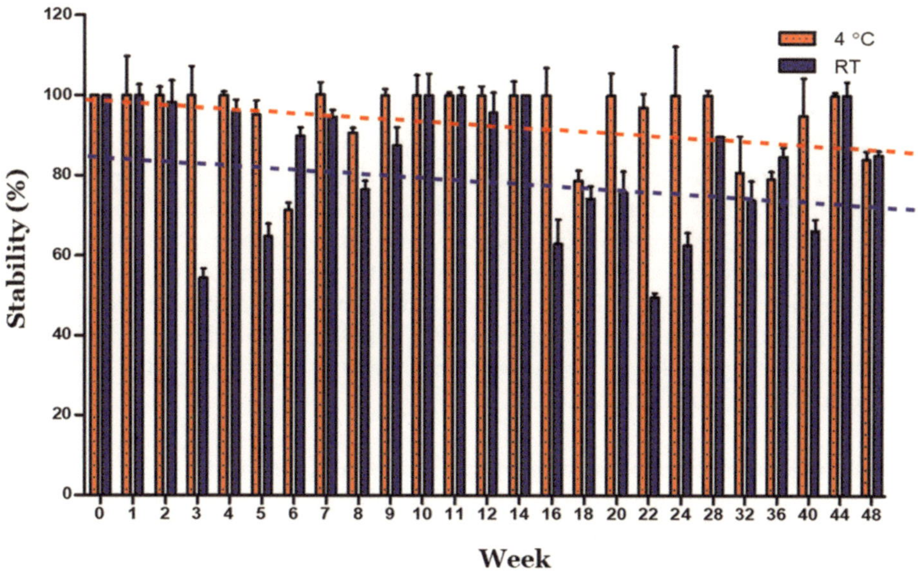

Figure 12.
Stability study of sensor strips with immobilized antibody against mycotoxin for a duration of 48 weeks stored in different temperatures (RT and 4°C).

4.4 Storage stability of the modified SPCEs

The stability of modified SPCEs with immobilized anti-AFB1 and anti-OTA antibodies was evaluated for the duration of 48 consecutive weeks. The sensor strips were stored without the presence of buffer at two storage temperatures: 4°C and room temperature (RT). This study is important to study the shelf-life of the functionalized strip in producing optimum results for long-term application. From **Figure 12**, it was found that storage in chilled environment showed higher antibody stability (90%) compared to ambient surroundings. The performance of the SPCEs started to decline below 90% stability after 14 weeks at 4°C and at the end of 48 weeks, the performance is still acceptable at 85% stability.

5. Electrochemical portable device as point-of-care

Portable electrochemical devices have a great potential as excellent tools for point-of-care testing technology. Detection of hazardous contaminants by means of portable electrochemical devices is a strategic approach for an efficient and rapid detection. With the complement of miniatured electrochemical cells, analysis can be conveniently performed on-site by nontechnical personnel. To facilitate on-site application and real-time detection of mycotoxins, a series of portable biosensor readers have been fabricated [39, 43, 44]. However, these readers, and even some commercially available handheld potentiostats, do not have latest features, particularly Internet-of-Things (IoT) integration. Recent trend in linking the portable device with IoT and cloud data storage has been made possible with the advancement of Fourth Industrial Revolution (IR 4.0).

In this study, an Android device is improvised as a universal portable electrochemical reader. A single unit of the portable reader is encased in a plastic box with dimensions of 16.8 cm (long) × 8.5 cm (wide) × 3.4 cm (height), powered by rechargeable lithium battery using USB cable through charger port. The reader which has a slot for SPCE attachment port consists of two main parts; an electrochemical transducer represented by a printed circuit board that converts signal from biology interactions reactions into readable and analyzable data; and a touch-screen Android-based device with developed electrochemical software application (**Figure 13**). In conventional electrochemical instrument, this unit is represented by a potentiostat. A Global Positioning System (GPS) tracking app is incorporated into the portable device, which allows users to pinpoint the real-time location of the field analysis performed. The handheld device has sufficient memory size of up to 32GB that can store over 100,000 measurements.

5.1 Interface for electrochemical portable reader and web server

The device interface is designed to be user-friendly, with the aim to allow users with minimal training to perform the testing themselves in the field. Authorized users will first need to log into the system before starting the measurement. This touch-screen device features drop-down menu, which is deemed convenient and practical for users' applications. To perform the testing, the user will need to insert the biosensor strip into the SPCE port on the reader, select the type of analyte (mycotoxins) to be tested, and press 'SCAN' to start the testing process. The interface for the Android-based device is shown in **Figure 14**. Once the scan is completed, the result is displayed on the device as concentration in ppb unit. The data then will be saved into the device, followed by the real-time transmission of the data together with GPS location and date/time to a central cloud server. The brightness of the screen can be adjusted; this feature is particularly important if the device is being used outdoor on the field. After

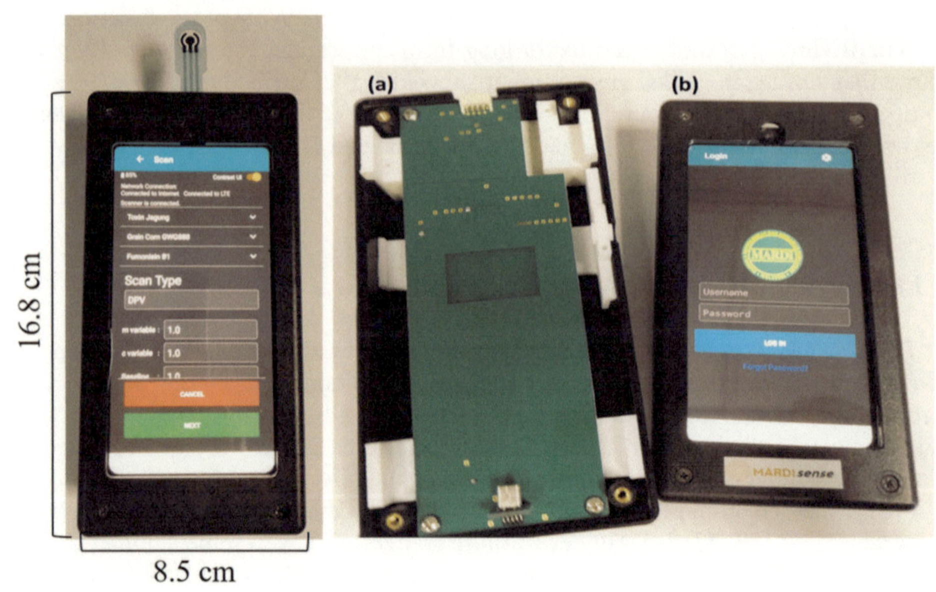

Figure 13.
An android-based portable electrochemical reader unit (16.8 x 8.5 x 3.4 cm) comprised of (a) electrochemical transducer and (b) touch-screen android-based device with developed electrochemical software.

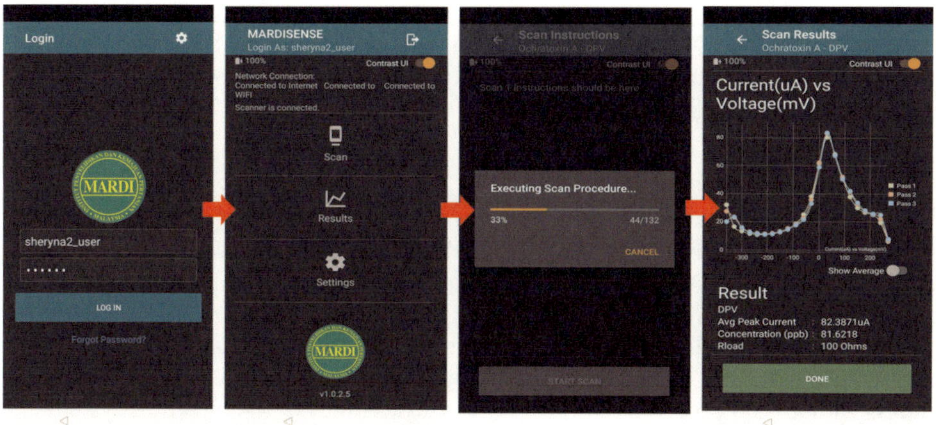

Figure 14.
Screen interface for portable reader application. User can select 'scan' to begin the analysis or 'results' to view the analyzed results. Results are displayed in form of graphs and readily calculated in concentrations (ppb).

a successful log-in, personnel can start a measurement, review results, or change the device setting.

All analyzed results on the portable readers will be stored on the cloud server with their data log details and pinned sample location. **Figure 15** displayed the web page for IoT server that acts as authorization, electrochemical setting, and data storage platform. The website can be accessed at http://mardisense.mardi.gov.my. Complete functions of each component were described in other publications [45].

5.2 Performance of portable reader as biosensing device

5.2.1 Sensor optimization on portable reader

Antibody concentration for both AFB1 and OTA was optimized at 0.1 mg/mL and was applied for cross-reaction study using a DPV measurement. The selectivity of immobilized in-house polyclonal antibody on the SPCEs was examined through cross-reactivity within the same toxins groups and other mycotoxins groups. As shown in

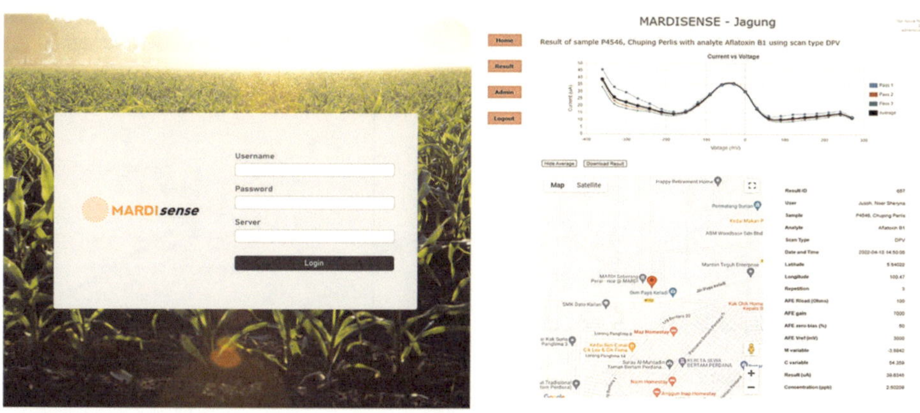

Figure 15.
Login interface of MARDIsense website (left); and page displaying details of the analyzed data logs (right).

Figure 16, the immuno-modified SPCEs were highly selective toward AFB1 and OTA, and demonstrated no cross-reaction with other mycotoxins. The cross-reactivity shown is less than 40%, hence the developed biosensor is still significantly sensitive and selective toward targeted toxins.

5.2.2 Sensitivity in sample matrix

The performance of developed sensor on the portable device was evaluated in undiluted and interference-free grain corn matrix. Standard curves for AFB1 and OTA are plotted as shown in **Figure 17**. Both AFB1 and OTA displayed an excellent linear correlation (R^2) of 0.9987 and 0.9836, respectively in a broad working range of 0–10 ppb. These indicate that the strategies of using the modified SPCEs on the portable reader are a success without compromising the sensitivity and accuracy of the biosensor system.

A recovery study was conducted for both AFB1 and OTA using spiked grain corn sample. This experiment is vital to evaluate the capability of the developed sensor to determine the presence of target analyte in sample. Prior to recovery study, the grain corns were first sterilized in 30% sodium hypochlorite (v/v) and autoclaved in ensuring a clean reference material for the spiking study. At concentrations of 10 ppb spiked mycotoxins, the recovery was found to be in the acceptable range of 89–96% (**Table 1**).

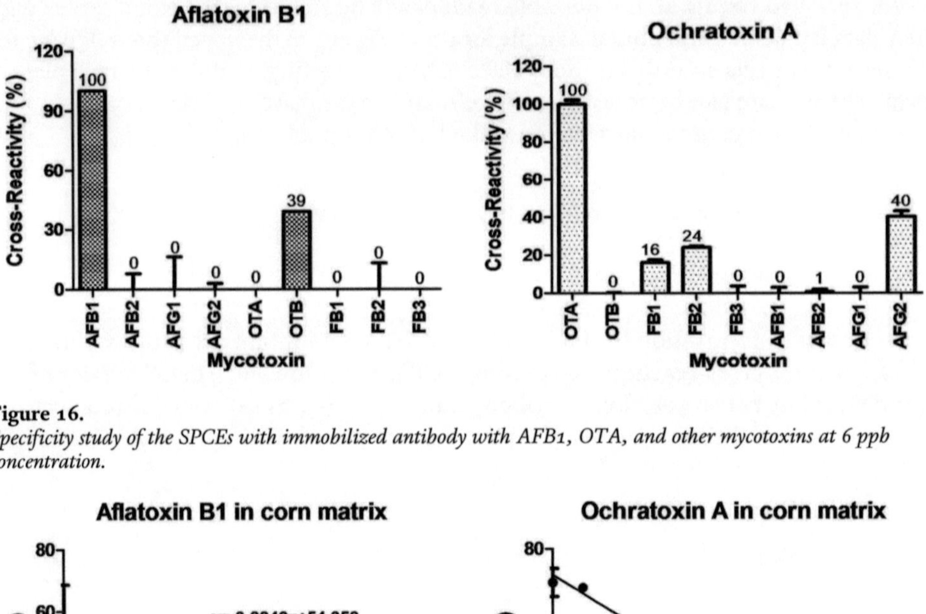

Figure 16.
Specificity study of the SPCEs with immobilized antibody with AFB1, OTA, and other mycotoxins at 6 ppb concentration.

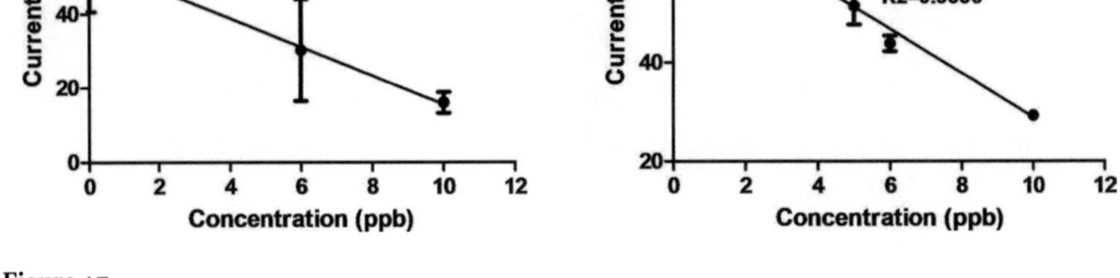

Figure 17.
Linear response from the portable device for a series of AFB1 (left) and OTA (right) concentrations with current reading in grain corn matrix.

DOI: http://dx.doi.org/10.5772/intechopen.111724

5.2.3 Development of immunosensing methods for mycotoxin determination in sample matrix

The sensitivity of AFB1 and OTA detections in grain corn samples on the portable device is studied and compared with electrochemical measurements on a laboratory potentiostat workstation and an immunoassay technique that is enzyme-linked immunosorbent assay (ELISA). For the latter, a direct competitive ELISA was applied and the antibody-antigen response was recorded from the absorbance reading. All of the three detection means described here employed the in-house polyclonal antibodies against AFB1 and OTA as their bioreceptors. The laboratory potentiostat was utilized for laboratory testing while the portable device was intended for on-field testing. Standard curves for different AFB1 and OTA concentrations were developed, and parameters from the standard curves were compared, namely correlation coefficient (R^2), limit of detection (LOD), and limit of quantitation (LOQ) values.

From **Table 2**, all immune-based methods exhibited high R^2 values in broad working range of 0–10 ppb except for ELISA for OTA, which has a working range between 0 and 50 ppb. Electrochemical sensing method, both on the electrochemical workstation and the IoT-portable reader, displayed good correlation values, which are comparable with those of ELISA method for AFB1 and OTA. The other important parameter of determining the sensitivity of a developed method is the limit of LOD and LOQ values. LOD value represents the lowest concentration of an analyte that can be detected using an instrument. Lower LOD value implied higher sensitivity of an instrument, thus proving the better performance in detecting the target analyte. The

	Spiked (ppb)	Found (ppb)	Recovery (%)
AFB1	10	8.86	89
OTA	10	9.60	96

Table 1.
Recovery study of 10 ppb AFB1 and OTA in spiked grain corn on portable-IoT device.

	Mycotoxins/Parameters	AFB1	OTA
ELISA	R^2	0.9756 (0–10 ppb)	0.9903 (0–50 ppb)
	LOD	1.90 ppb	2.70 ppb
	LOQ	5.77 ppb	8.10 ppb
Electrochemical workstation	R^2	0.971 (0–10 ppb)	0.9685 (0–10 ppb)
	LOD	2.54 ppb	3.31 ppb
	LOQ	7.68 ppb	10.03 ppb
IoT-Portable reader	R^2	0.9987 (0–10 ppb)	0.9836 (0–10 ppb)
	LOD	0.84 ppb	1.72 ppb
	LOQ	2.55 ppb	5.22 ppb

Table 2.
Comparison of the performance for immuno-based detection methods for AFB1 and OTA in grain corn matrix.

MRL for both AFB1 and OTA in feed is 5 ppb and the calculated LOD values for both mycotoxins were much lower than the MRL for all immuno-based methods. In particular, the developed portable device recorded the lowest LOD and LOQ values of all for both AFB1 and OTA, signifying that the device was applicable for mycotoxins on-site testing.

5.3 On-field application of portable reader

In-situ determination of AFB1 was conducted at a cornfield in Perlis, northern region state of Malaysia (GPS coordinate 6.59835, 100.283) in March (pre-harvest season) and April 2022 (harvest season). Preharvest testing was carried out on 90-day-old grain corn for primary detection of AFB1 while harvest testing was carried out on 117-day-old grain corn. Samples (grain corn cob) were taken from five sampling plots, extracted using a simple sample extraction, and measured on-field (**Figure 18**). All samples showed detected AB1 concentration below the permitted level, which is 5 ppb [45].

5.4 Validation study with instrumentation method

Besides samples in Perlis, the developed biosensor system with the portable device was also tested on several other grain corn samples from Bachok and Seberang Perai, and used for AFB1 monitoring in stored grain corn with different packaging materials. A total of twenty-grain corn samples were analyzed with the portable reader and validated with HPLC-FL post-column method and UPLC-FLR. Correlation between the analyzed data and the instrumentation is tabulated in **Table 3**. LOD for the electrochemical portable reader sensor is 0.84 ppb while LOD for HPLC is 2.5 ppb. Data for HPLC method is presented by LOD reading, and it is indicative of either the AFB1 is not detected by the system or AFB1 concentration is below 2.5 ppb.

Figure 18.
On-site application for AFB1 detection in grain corn using the modified SPCEs and portable reader.

	[a]**Portable device**	[b]**HPLC**
Sample 1	0.96	<LOD
Sample 2	[c]n.d.	<LOD
Sample 3	n.d.	<LOD
Sample 4	n.d.	<LOD
Sample 5	n.d.	<LOD
Sample 6	n.d.	<LOD
Sample 7	n.d.	<LOD
Sample 8	1.25	<LOQ
Sample 9	n.d.	<LOD
Sample 10	1.51	<LOD
Sample 11	1.25	<LOD
Sample 12	2.34	<LOD
Sample 13	3.42	<LOD
Sample 14	n.d.	<LOD
Sample 15	4.32	2.85
Sample 16	3.44	<LOD
	Portable device	[d]UPLC-FLR
Sample 17	n.d.	n.d.
Sample 18	n.d.	n.d.
Sample 19	<LOD	n.d.
Sample 20	2.13	n.d.

[a]LOD electrochemical portable device = 0.84 ppb.[b]LOD HPLC = 2.5 ppb.[c]n.d. = not detected.[d]LOQ UPLC-FLR = 1 ppb.

Table 3.
Detection of AFBI in grain corn sample on the portable reader and its validation study using instrumentation methods (HPLC and UPLC-FLR).

Of the 20 samples, ten samples were not contaminated by AFB1 as determined by the portable reader. Nine samples were detected with low AFB1 concentrations lower than the MRL permitted (5 ppb). Sample 15 recorded 4.42 ppb of AFB1 and was confirmed by the HPLC with 2.85 ppb. The discrepancies detection for validation with the UPLC-FLR method could be attributed to different sampling batches and duration of time for both experiments conducted, which affect the moisture content in the samples.

In general, the strategies of biosensor miniaturization for point-of-care using the modified SPCEs on the portable reader are successful without compromising the sensitivity and accuracy of the biosensor system. The portable reader offers a straightforward and direct application of AFB1 and OTA detection in grain corn with simple sample extraction as opposed to conventional methods. Furthermore, biosensor with portable reader allows the test to be performed rapidly on-site without the need to bring back the samples to the laboratory where the transportation and logistics factors may also contribute to the discrepancies in the analyzed results. The portable reader and miniaturized biosensor system described here can be used for mobile lab applications. Although the studies presented here highlighted mainly grain corn, nevertheless the application of this developed system can be widened to other food commodities, particularly peanuts and rice for AFB1; and grains and coffee for OTA, to name a few. Moving forward, another point to be considered is the development of multi-mycotoxins detection on a dual or multiple working electrode [46].

6. Conclusions

A portable electrochemical device integrated with IoT system has been successfully designed for rapid and *in-situ* detection of *Aspergillus* spp. mycotoxins. Polyclonal antibody against AFB1 and OTA that were produced and purified in-house showed excellent sensitivity and selectivity toward the targeted mycotoxins. The antibodies are then further utilized and immobilized on modified SPCEs for the development of immunosensor system for mycotoxins detection. Using grain corn as matrix sample model, electrochemical measurements achieved by DPV for AFB1 and OTA standard curves development on the portable IoT device are superior to the ELISA method and laboratory electrochemical workstation. Good linearity was obtained in working range of 0–10 ppb (MRL 5 ppb) with excellent LOD and LOQ values. *In-situ* analysis of AB1 has been successfully conducted at a local cornfield in Perlis, northern Malaysia, using the portable device integrated with IoT as point-of-care tool. Analysis of 20 samples with the portable reader correlated well with HPLC and UPLC-FLR methods. In general, biosensors posed major advantages over the conventional method for mycotoxins detection in terms of assay duration, detection limit, and portability. With IoT integration and web server that can store data logs and pinpoint sampling locations, the portable device posed an excellent tool for mycotoxins routine monitoring in near future.

Acknowledgements

The authors would like to thank the Ministry of Agriculture and Food Industry of Malaysia for the financial support under research fund KRB 167. Gratitude also goes to other team members involved in this study and Mr. Hashim Ab Rahim from Persada Bumi Enterprise for the sampling plot at the cornfield.

Conflict of interest

The authors declare no conflict of interest.

Fabrication of portable device

Electrochemical portable devices (Version 1) with cloud server integration were fabricated and developed by Biogenes Technologies Sdn. Bhd., Malaysia.

Author details

Nur Azura Mohd Said*, Noor Sheryna Jusoh, Norhafniza Awaludin, Mohammad Rejab Ismail, Noor Fadilah Mohd Bakri, Lily Suhaida Mohd Sojak and Faridah Salam
Biotechnology and Nanotechnology Research Centre, Malaysian Agricultural Research and Development Institute (MARDI), Selangor, Malaysia

*Address all correspondence to: nazurams@mardi.gov.my

References

[1] Reddy KR, Abbas HK, Abel CA, Shier WT, Oliveira CA, Raghavender CR. Mycotoxin contamination of commercially important agricultural commodities. Toxin Reviews. 2009;**28**(2–3):154-168

[2] Rao SB, Prasad KS, Rajendran D. Recent advances in amelioration of anti-nutritional factors in livestock feedstuffs. In: Animal Nutrition & Reproductive Physiology (Recent Concepts). Vol. 13. Delhi, India: Satish Serial Publishing House; 2013. pp. 655-678

[3] Oliveira M, Vasconcelos V. Occurrence of mycotoxins in fish feed and its effects: A review. Toxins. 2020; **12**(3):160

[4] Gonçalves RA, Schatzmayr D, Albalat A, Mackenzie S. Mycotoxins in aquaculture: Feed and food. Reviews in Aquaculture. 2020;**12**(1):145-175

[5] da Rocha ME, Freire FD, Maia FE, Guedes MI, Rondina D. Mycotoxins and their effects on human and animal health. Food Control. 2014;**36**(1):159-165

[6] Udomkun P, Wiredu AN, Nagle M, Bandyopadhyay R, Müller J, Vanlauwe B. Mycotoxins in sub-Saharan Africa: Present situation, socio-economic impact, awareness, and outlook. Food Control. 2017;**1**(72):110-122

[7] Jiang Y, Ogunade IM, Vyas D, Adesogan AT. Aflatoxin in dairy cows: Toxicity, occurrence in feedstuffs and milk and dietary mitigation strategies. Toxins. 2021;**13**(4):283

[8] World Health Organization, International Agency for Research on Cancer. Some naturally occurring substances: Food items and constituents, heterocyclic aromatic amines and mycotoxins. In: IARC Monographs on the Evaluation of the Carcinogenic Risk of Chemicals to Humans. Lyon, France: World Health Organization, International Agency for Research on Cancer; 1993. p. 56

[9] Lakkireddy K, Kasturi K, Rao KRSS. Aflatoxins in food and feed: The science of safe food. Practice. 2014;**3**:5

[10] BIOMIN Mycotoxin Survey Q2 2021 Results [Internet]. 2022. Available from: https://www.biomin.net/science-hub/biomin-mycotoxin-survey-q2-2021-results/ [Accessed: April 1, 2023]

[11] Kumar P, Mahato DK, Sharma B, Borah R, Haque S, Mahmud MC, et al. Ochratoxins in food and feed: Occurrence and its impact on human health and management strategies. Toxicon. 2020;**1**(187):151-162

[12] Cui X, Jin M, Du P, Chen G, Zhang C, Zhang Y, et al. Development of immunoassays for multi-residue detection of small molecule compounds. Food and Agricultural Immunology. 2018;**29**(1):638-652

[13] Laura A, Claudio B, Cristina G, Gianfranco G. A lateral flow immunoassay for measuring ochratoxin a: Development of a single system for maize, wheat and durum wheat. Food Control. 2011;**22**(12):1965-1970

[14] Senthilkumar T, Jayas DS, White ND, Fields PG, Gräfenhan T. Detection of fungal infection and Ochratoxin a contamination in stored barley using near-infrared hyperspectral imaging. Biosystems Engineering. 2016; **1**(147):162-173

[15] Senthilkumar T, Jayas DS, White ND, Fields PG, Gräfenhan T.

Detection of fungal infection and Ochratoxin a contamination in stored wheat using near-infrared hyperspectral imaging. Journal of Stored Products Research. 2016;**1**(65):30-39

[16] Wang B, Deng J, Jiang H. Markov transition field combined with convolutional neural network improved the predictive performance of near-infrared spectroscopy models for determination of aflatoxin B1 in maize. Food. 2022;**11**(15):2210

[17] Zhou Q, Huang W, Tian X. Feature wavelength selection based on the combination of image and Spectrum for aflatoxin B1 concentration classification in single maize kernels. Agriculture. 2022;**12**(3):385

[18] Bertani FR, Businaro L, Gambacorta L, Mencattini A, Brenda D, Di Giuseppe D, et al. Optical detection of aflatoxins B in grained almonds using fluorescence spectroscopy and machine learning algorithms. Food Control. 2020; **1**(112):107073

[19] Wang X, Bouzembrak Y, Oude Lansink AG, Van der Fels-Klerx HJ. Designing a monitoring program for aflatoxin B1 in feed products using machine learning. npj Science of Food. 2022;**6**(1):40

[20] Clark LC Jr, Lyons C. Electrode systems for continuous monitoring in cardiovascular surgery. Annals of the New York Academy of sciences. 1962; **102**(1):29-45

[21] Kirsch J, Siltanen C, Zhou Q, Revzin A, Simonian A. Biosensor technology: Recent advances in threat agent detection and medicine. Chemical Society Reviews. 2013;**42**(22):8733-8768

[22] Alhadrami HA. Biosensors: Classifications, medical applications, and future prospective. Biotechnology and Applied Biochemistry. 2018;**65**(3): 497-508

[23] Olejnik B, Kozioł A, Brzozowska E, Ferens-Sieczkowska M. Application of selected biosensor techniques in clinical diagnostics. Expert review of molecular diagnostics. 2021;**21**(9):925-937

[24] Yasmin J, Ahmed MR, Cho BK. Biosensors and their applications in food safety: A review. Journal of Biosystems Engineering. 2016;**41**(3):240-254

[25] Zhang J, Huang H, Song G, Huang K, Luo Y, Liu Q, et al. Intelligent biosensing strategies for rapid detection in food safety: A review. Biosensors and Bioelectronics. 2022;**17**: 114003

[26] Adetunji CO, Nwankwo W, Ukhurebor KE, Olayinka AS, Makinde AS. Application of biosensor for the identification of various pathogens and pests mitigating against the agricultural production: Recent advances. In: Biosensors in Agriculture: Recent Trends and Future Perspectives. Cham: Springer; 2021. pp. 169-189

[27] Dyussembayev K, Sambasivam P, Bar I, Brownlie JC, Shiddiky MJ, Ford R. Biosensor technologies for early detection and quantification of plant pathogens. Frontiers in Chemistry. 2021; **2**(9):636245

[28] Griesche C, Baeumner AJ. Biosensors to support sustainable agriculture and food safety. TrAC Trends in Analytical Chemistry. 2020;**1**(128):115906

[29] Bahadır EB, Sezgintürk MK. Applications of commercial biosensors in clinical, food, environmental, and biothreat/biowarfare analyses. Analytical Biochemistry. 2015;**1**(478): 107-120

[30] Collings AF, Caruso F. Biosensors: recent advances. Reports on Progress in Physics. 1997;**60**(11):1397

[31] Mohd Said NA. Electrochemical Biosensor Based on Microfabricated Electrode Arrays for Life Sciences Applications [Thesis]. Ireland: University College Cork; 2014

[32] Curulli A. Electrochemical biosensors in food safety: Challenges and perspectives. Molecules. 2021;**26**(10): 2940

[33] Wang K, Lin X, Zhang M, Li Y, Luo C, Wu J. Review of electrochemical biosensors for food safety detection. Biosensors. 2022;**12**(11):959

[34] Mishra GK, Barfidokht A, Tehrani F, Mishra RK. Food safety analysis using electrochemical biosensors. Food. 2018; **7**(9):141

[35] Jiang C, Lan L, Yao Y, Zhao F, Ping J. Recent progress in application of nanomaterial-enabled biosensors for ochratoxin a detection. TrAC Trends in Analytical Chemistry. 2018;**1**(102): 236-249

[36] Parihar A, Choudhary NK, Sharma P, Khan R. MXene-based aptasensor for the detection of aflatoxin in food and agricultural products. Environmental Pollution. 2022;**21**: 120695

[37] Leenaars M, Hendriksen CF. Critical steps in the production of polyclonal and monoclonal antibodies: Evaluation and recommendations. ILAR journal. 2005; **46**(3):269-279

[38] Lee NA, Wang S, Allan RD, Kennedy IR. A rapid aflatoxin B1 ELISA: Development and validation with reduced matrix effects for peanuts, corn, pistachio, and soybeans. Journal of agricultural and food chemistry. 2004; **52**(10):2746-2755

[39] Umapathi R, Ghoreishian SM, Rani GM, Cho Y, Huh YS. Emerging trends in the development of electrochemical devices for the on-site detection of food contaminants. ECS Sensors Plus. 2022;**1**(4):044601

[40] Yamanaka K, Vestergaard MD, Tamiya E. Printable electrochemical biosensors: A focus on screen-printed electrodes and their application. Sensors. 2016;**16**(10):1761

[41] Taleat Z, Khoshroo A, Mazloum-Ardakani M. Screen-printed electrodes for biosensing: A review (2008–2013). Microchimica Acta. 2014;**181**:865-891

[42] Jusoh NS, Awaludin N, Salam F, Kadir AA, Said NA. Label-free electrochemical Immunosensor development for mycotoxins detection in grain corn. Malaysian Journal of Analytical Sciences. 2022;**26**(6):1205-1215

[43] Eyvazi S, Baradaran B, Mokhtarzadeh A, de la Guardia M. Recent advances on development of portable biosensors for monitoring of biological contaminants in foods. Trends in Food Science & Technology. 2021; **1**(114):712-721

[44] Ogurtsov VI, Twomey K. A Portable chemical detection system with anti-body biosensor for impedance based monitoring of T2-mycotoxin bioterrorism agents. In: Biomedical Engineering Systems and Technologies: 10th International Joint Conference (BIOSTEC 2017): 21-23 February 2017; Porto, Portugal. New York: Springer International Publishing; 2018. pp. 45-73

[45] Jusoh NS, Said NA, Salam F, Awaludin N, Kadir AA, Ismail MR. An integrated handheld biosensing device

with IoT system for in-situ and rapid detection of multi-mycotoxins. In: 2022 4th International Conference on Smart Sensors and Application (ICSSA); 26-28 July 2022; Kuala Lumpur. New York: IEEE; 2022. pp. 11-16

[46] Jahangiri-Dehaghani F, Zare HR, Shekari Z. Simultaneous Measurement of Ochratoxin a and Aflatoxin B1 Using a Duplexed-Electrochemical Aptasensor Based on Carbon Nanodots Decorated with Gold Nanoparticles and Two Redox Probes Hemin@ Hkust-1 and Ferrocene@ Hkust-1. SSRN. 2023. DOI: 10.2139/ssrn.4395228

Chapter 9

The Menace of Aflatoxin: Understanding the Effects of Contamination by *Aspergillus Species* on Crops and Human Health and Advancements in Managing These Toxic Metabolites

Amir Afzal, Sairah Syed, Rafiq Ahmad, Muhammad Zeeshan and Ghulam Nabi

Abstract

Food security and safety are essential global issues that require collaboration from governments, private industry, and individuals to ensure there is enough safe and nutritious food to meet the needs of a growing population. The three main elements impacting food security and safety are the availability of food, access to safe food, and the utilization of food for a healthy life. Aflatoxins, harmful mycotoxins produced by certain fungi, damage a significant proportion of the world's food supply, which is a factor in food insecurity. Effective strategies to prevent and manage aflatoxin contamination in crops include promoting sustainable and diversified agricultural practices, improving crop management, post-harvest handling and storage, and strict regulation and monitoring of food quality and safety. To date, there have been 20 different types of aflatoxins identified, with B_1, B_2, G_1, and G_2 being the most prevalent and dangerous. To mitigate the impact of aflatoxins, it is important to understand the mechanisms of contamination, the impact of aflatoxins, and the management strategies that can be employed to reduce contamination. An updated review on aflatoxin contamination, its impact and management strategies can provide valuable information for researchers, policymakers, and food safety professionals.

Keywords: food safety, aflatoxins, aflatoxin contamination, *Aspergillus* spp., toxic metabolites

1. Introduction

Food security and safety are indeed global issues that are becoming increasingly important as the world's population continues to grow. Food security refers to the ability of individuals and communities to access sufficient, safe, and nutritious food to meet their dietary needs and preferences. With a growing population, it is essential to ensure that there is enough food available to meet the nutritional requirements of everyone. Food safety, on the other hand, refers to the measures taken to ensure that food is free from harmful contaminants and pathogens that could cause illness or disease. This includes everything from ensuring that food is properly stored and handled to conducting regular inspections of food processing facilities and enforcing strict food safety regulations. Ensuring both food security and food safety requires a coordinated effort from governments, private industry, and individuals. This includes investing in sustainable agriculture and food production practices, supporting local food systems, and promoting nutrition education and awareness. Ultimately, ensuring food security and safety is critical for the health and well-being of individuals and communities around the world, and it will require ongoing commitment and collaboration from all stakeholders to achieve this goal. Food security and safety are among the top priorities in today's world with a growing population. These issues are primarily influenced by three critical elements: (1) sufficient food availability, (2) accessibility to safe food, and (3) utilization of food with regard to its quality, nutritional value, and cultural significance for a healthy lifestyle. Having enough food is essential to ensure individuals have access to enough sustenance to meet their daily energy requirements and maintain good health. However, food security goes beyond mere availability and encompasses the quality and safety of the food as well. Access to safe food is also a concern, particularly for those who may not have access due to poverty, limited resources, or unavailability [1]. In some areas, food may become contaminated due to improper storage, handling, or transportation practices, making it necessary to implement measures to ensure proper food handling and storage. Finally, utilizing food in terms of quality, nutrition, and cultural significance is important for a healthy life. Food provides necessary nutrients for growth and body maintenance, and also has a significant cultural and social impact [2]. A balanced and varied diet is necessary to maintain good health and prevent chronic diseases. In conclusion, addressing these crucial elements of food security and safety is imperative for the overall well-being of the global population. If any of these components fail, it results in food insecurity and malnutrition that has a detrimental impact on human health and the social and economic well-being of society. Furthermore, contamination of food and feed by mycotoxins is a significant contributor to the problem of food insecurity [3]. Aflatoxins are a type of mycotoxin primarily produced by the fungi *Aspergillus flavus* and *A. parasiticus* and have been extensively studied [4]. In addition to the human and animal health risks associated with aflatoxin contamination, it can also have significant economic impacts. Aflatoxin-contaminated crops may be rejected by food processors and retailers, resulting in reduced market value and financial losses for farmers. In some cases, crops may need to be destroyed entirely, which can result in even greater economic losses. The contamination of a diverse range of foods and feeds with AF can lead to economic losses, and various factors such as season, post-harvest and management practices, food type, and geographic location can contribute to this contamination [5]. This review serves as a valuable resource for researchers as it provides crucial information that can be used to devise effective mitigating strategies.

According to the Centers for Disease Control and Prevention (CDC), mycotoxin exposure is a chronic issue affecting an estimated 4.5 billion people [6]. However, a recent study suggests that the global occurrence of mycotoxin contamination in crops is between 60 and 80% [7].

2. Aflatoxin-producing species of *Aspergillus* and their diversity

Aflatoxins, which are primarily produced by *Aspergillus flavus* and A. parasiticus, are the mycotoxins known to be produced by species of *Aspergillus*. However, other species such as *A. nomius, A. pseudotamarii, A. parvisclerotigenus,* and *A. bombycis* from section *Flavi*, *A. ochraceoroseus* and *A. rambellii* from section *Ochraceorosei*, as well as *Emericella astellata* and *Epipleoneura venezuelensis* from the *Nidulatans*, have also been identified as producers of aflatoxins [8]. There are multiple types of aflatoxins that have been documented, and their presence in crops and food is a major global concern, particularly for crops of economic significance (**Figure 1**). Aflatoxins are toxic compounds produced by certain species of fungi, including *Aspergillus flavus* and *A. parasiticus* [10].

The genus *Aspergillus* is comprised of four subgenera and a total of 339 species [11]. The *Aspergillus* genus is a diverse group of fungi, which includes over 200 species [12]. The agriculturally important species of *Aspergillus* that produce aflatoxin belong to the *Flavi* section [13]. *A. flavus* and *A. parasiticus* are known to contaminate a variety of crops including maize, peanuts, cottonseed, tree nuts, and spices, among others [14]. There are two different types of *A. flavus*, a fungus that creates aflatoxins, based on the size of their sclerotia: L and S morphotypes. The L morphotype produces many spores and a range of aflatoxin levels, but only a few large sclerotia (>400 μm). The S morphotype produces fewer spores, but consistently high levels of aflatoxins, and many small sclerotia (<400 μm) [15].

3. Classification and types of aflatoxins

The names of these aflatoxins are derived from their property of absorbing and emitting light at distinct wavelengths. Aflatoxins B_1 and B_2 fluoresce blue when subjected to ultraviolet light with a wavelength of 425 nm, while aflatoxins G_1 and G_2 fluoresce green when exposed to ultraviolet light with a wavelength of 540 nm [5]. Aflatoxins are difuranocoumarins that can be divided into two categories based on their chemical structure: the difurocoumarocyclopentenone series, which includes AFB_1, AFB_2, AFB_2A, AFM_1, AFM_2, AFM_2A, and aflatoxicol; and the difurocoumarolactone series, which includes AFG_1, AFG_2, AFG_2A, $AFGM_1$, $AFGM_2$, $AFGM_2A$, and

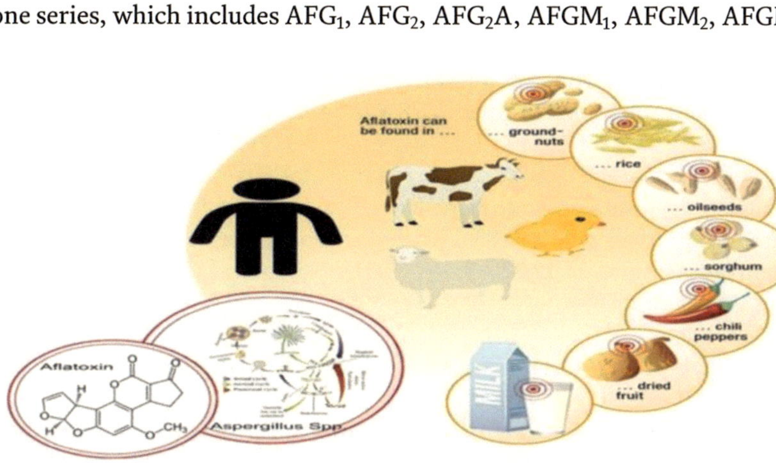

Figure 1.
Contamination of aflatoxins in a wide range of regional crops and food commodities [9].

AFB_3 [16]. More than 20 distinct types of aflatoxins have been identified to date [17]. However, the most commonly and widely known aflatoxins are aflatoxins B_1, B_2, G_1, G_2, M_1, M_2, aflatoxicol, and aflatoxin Q_1 [18].

4. Environmental conditions favoring aflatoxin production on crops

The growth of these fungi is commonly observed in warm and humid conditions, which are prevalent in tropical and subtropical regions [19]. They are commonly found on crops such as corn, peanuts, cottonseed, and tree nuts, which are stored in warm and humid conditions for extended periods of time. This provides an ideal environment for the growth and spread of aflatoxin-producing fungi, leading to contamination of these crops and the potential health hazards and economic losses that come with it [20].

5. The carcinogenic properties of aflatoxins B_1, B_2, G_1, and G_2

Of all the types of aflatoxins, aflatoxins B_1, B_2, G_1, and G_2 have received the most extensive research attention and are the most commonly found, and they have been shown to have the most significant impact on human and animal health [21, 22]. These are considered to be more important due to their widespread prevalence in food [5]. Aflatoxins B_1, B_2, G_1, and G_2 are four major types of aflatoxins that are classified as Group 1 carcinogens. Among them, the most toxic and commonly found in crops is aflatoxin B_1 (AFB_1) [23]. Moreover, aflatoxin B_1 can bind to DNA and modify its structure, resulting in genotoxic effects [24]. Different species of *Aspergillus* fungi produce different types of aflatoxins. Aflatoxins B_1, B_2, G_1, and G_2 are produced by *Aspergillus bombycis*, *A. nomius*, *A. parasiticus*, *A. parvisclerotigenus*, *A. pseudocaelatus*, *A. minisclerotigenes*, and *A. arachidicola*. However, species such as *A. flavus*, *A. ochraceoroseus*, and *A. rambellii* only produce aflatoxin B_1 and B_2, while others such as *Aspergillus pseudonomius, A. pseudotamarii, Emericella astellata*, *E. olivicola*, and *Epipleoneura venezuelensis* produce only aflatoxin B_1 [25]. The occurrence of aflatoxins was observed in the following decreasing order: $AFG_2 > AFG_1 > AFB_2 > AFB_1$, where: AFG_1: Aflatoxin G_1, AFG_2: Aflatoxin G_2, AFB_1: Aflatoxin B_1, AFB_2: Aflatoxin B_2. Note that the ">"sign indicates "greater than" and is used to show the decreasing order of occurrence, with AFG2 being the most prevalent and AFB_1 being the least prevalent with concentration ranges of 0.78 ± 0.04–234.73 ± 3.8 μg/kg, 0.47 ± 0.03–21.6 ± 0.33 μg/kg, 1.01 ± 0.05–13.75 ± 1.2 μg/kg, and 0.66 ± 0.06–5.51 ± 0.26 μg/kg, respectively. Of the 100 samples analyzed for total aflatoxins (total AFs), 68 (68%) exceeded the limits set by the EC, with concentration ranges of 4.98 ± 0.6–445.01 ± 8.9 μg/kg. Similarly, 58 (58%) of the samples exceeded the limits set by GSA, with concentration ranges of 12.12 ± 1.4–445.01 ± 8.9 μg/kg [26].

6. Understanding the health effects of aflatoxin exposure

Aflatoxin contamination is a significant public health concern, particularly in developing countries where food safety regulations may be less stringent and where poverty and malnutrition can exacerbate the health effects of exposure [27]. The European Commission and the U.S. Food and Drug Administration have set a maximum limit of 20 ppb (parts per billion) of aflatoxins in food and feed products for human and animal consumption to help prevent these health hazards and economic losses. Aflatoxins are both carcinogenic and mutagenic in nature and can result in aflatoxicosis in humans and animals [5]. Recently, a form of pulmonary aspergillosis has been linked to the coronavirus disease (COVID-19). It has been established that patients with compromised immune systems are more likely to experience severe cases of COVID-19 when complicated by pulmonary aspergillosis. To date, there have been 20 reported cases of COVID-19 associated pulmonary aspergillosis (CAPA) worldwide [28]. Recently, a correlation between the coronavirus disease (COVID-19) and pulmonary aspergillosis has also been reported, with 20 cases of coronavirus disease-associated pulmonary aspergillosis (CAPA) recorded globally. It is noted that the severity of COVID-19 symptoms is increased in immunocompromised patients with pulmonary aspergillosis [29]. Given their widespread presence, around 4.5 billion people worldwide are estimated to be exposed to aflatoxin contamination [30]. The adverse effects of aflatoxins on living organisms have led to the setting of a maximum limit of 20 ppb (parts per billion) for aflatoxins in food and feed products intended for human and animal consumption by both the European Commission and the U.S. Food and Drug Administration [31]. The adverse health effects of aflatoxin contamination can include acute toxicity, immune suppression, liver damage, and an increased risk of developing liver cancer. Aflatoxin contamination can have a range of adverse health effects in both humans and animals, and these effects can vary depending on the level and duration of exposure. The most significant health effects of aflatoxin contamination include the following.

6.1 The adverse health effects of aflatoxin contamination and the importance of prevention and control measures

High levels of aflatoxin exposure can cause acute toxicity, which can result in a range of symptoms, such as vomiting, abdominal pain, convulsions, coma, and even death ([32] Available from: https://www.ncbi.nlm.nih.gov/books/NBK557781/). These symptoms occur due to the liver's inability to detoxify aflatoxins effectively, leading to a buildup of toxic metabolites in the body. When consumed, aflatoxins are primarily metabolized in the liver, where they can cause damage to liver cells and impair liver function. The liver is responsible for breaking down and eliminating many toxins, including aflatoxins, from the body. However, in cases of high-level exposure, the liver may not be able to metabolize and eliminate the aflatoxins effectively, leading to a buildup of toxic metabolites in the body [33]. The severity of the symptoms of acute toxicity depends on several factors, including the level and duration of aflatoxin exposure, as well as the individual's age, nutritional status, and overall health. Young children and people with weakened immune systems or liver disease are particularly vulnerable to the adverse effects of aflatoxin exposure, as their bodies may not be able to detoxify the aflatoxins effectively. In addition, malnutrition can exacerbate the health effects of aflatoxin exposure, as it can weaken the body's ability to fight off infections and other health problems. This can lead to an increased risk of complications and death in cases of acute aflatoxin toxicity. Overall, the adverse health effects of aflatoxin contamination, including acute toxicity, can have significant impacts on public health and food safety, particularly in developing countries where food safety regulations may be less stringent, and where poverty and malnutrition can exacerbate the health effects of exposure. Implementing appropriate prevention and control measures, regular monitoring and testing, and appropriate food safety regulations, as well as promoting public awareness and education about the risks of aflatoxin exposure, are critical to preventing the adverse health effects of aflatoxin contamination.

6.2 Aflatoxin exposure and the increased risk of infections and diseases

Aflatoxins are known to have immunosuppressive effects, which means that they can impair the body's immune system and its ability to fight off infections and other diseases. A weakened immune system can increase the risk of various health problems, including bacterial and viral infections, as well as chronic illnesses such as cancer. The immunosuppressive effects of aflatoxins can occur through various mechanisms, including the disruption of immune cell function and the impairment of cytokine production. Aflatoxins have been shown to reduce the production of cytokines, which are proteins that regulate the immune response and play a critical role in fighting infections and diseases. In addition, aflatoxins can damage the liver, which can impair its ability to produce proteins that are important for immune function. The immunosuppressive effects of aflatoxins are particularly concerning for individuals who are already immunocompromised, such as those with HIV/AIDS, cancer, or other chronic illnesses. For these individuals, exposure to aflatoxins can further weaken their immune system, making them more susceptible to infections and other health problems. Coulombe [34] described the various routes of exposure to mycotoxins, such as ingestion, inhalation, and dermal contact. He discussed the factors that can influence the toxic effects of mycotoxins, such as dose, duration of exposure, and the age and health status of the individual.

6.3 Aflatoxins as carcinogens: understanding the link to liver cancer

Aflatoxins are toxic compounds that are primarily metabolized by the liver. When ingested, they are absorbed from the gastrointestinal tract and transported to the liver, where they are metabolized into a range of byproducts, some of which are highly reactive and can cause damage to liver cells. The liver plays a critical role in the detoxification of aflatoxins and other harmful substances, but when exposed to high levels of aflatoxins, the liver can become overwhelmed, and its ability to metabolize and eliminate the toxins can be impaired. This can lead to the accumulation of aflatoxin metabolites in the liver and other tissues, which can cause liver damage and even liver failure. Liver damage caused by aflatoxin exposure can manifest in various ways, including acute and chronic forms. Acute liver damage caused by aflatoxin exposure is characterized by rapid onset and is associated with symptoms such as jaundice, abdominal pain, and liver failure. Chronic liver damage caused by aflatoxin exposure, on the other hand, is associated with long-term exposure to lower levels of the toxin and can result in the development of liver cirrhosis or liver cancer. Individuals who are at higher risk of liver damage caused by aflatoxin exposure include those who consume a diet high in aflatoxin-contaminated foods, as well as those who have pre-existing liver disease or other risk factors that compromise liver function [35].

HCC is a primary liver cancer that arises from the hepatocytes, which are the main functional cells of the liver. HCC is a major global health problem, and its incidence has been increasing in many countries worldwide. The risk factors for HCC, which include chronic infection with hepatitis B virus (HBV) or hepatitis C virus (HCV), alcohol consumption, obesity, and exposure to aflatoxin. The prevalence of these risk factors varies in different regions of the world and among different populations [36].

Chronic exposure to aflatoxins has been shown to increase the risk of developing liver cancer, a type of cancer that affects the liver [30]. Aflatoxins are naturally occurring toxins produced by certain molds, primarily *Aspergillus flavus* and *Aspergillus parasiticus*, which can contaminate food and feed crops, particularly those that are stored in warm and humid conditions. When contaminated crops are consumed by humans or animals, aflatoxins can enter the bloodstream and be transported to the liver, where they can cause DNA mutations and other cellular damage that can lead to the development of cancerous cells.

Aflatoxins are classified as Group 1 carcinogens by the International Agency for Research on Cancer (IARC), which means that there is sufficient evidence to suggest that they are carcinogenic to humans. Aflatoxin exposure is a significant risk factor for liver cancer in areas where food and feed contamination are common, such as sub-Saharan Africa, Southeast Asia, and parts of South America. The liver is the primary organ responsible for detoxifying aflatoxins, but chronic exposure to high levels of aflatoxins can overwhelm the liver's capacity to metabolize and eliminate the toxins, leading to the accumulation of DNA-damaging metabolites and increased risk of liver cancer [37]. The risk of developing liver cancer from aflatoxin exposure is further increased in individuals with pre-existing liver disease, such as viral hepatitis B or C, as well as those who consume alcohol or have a weakened immune system. Aflatoxins are highly toxic and carcinogenic to both humans and animals, and exposure to these toxins can lead to liver damage, immune system suppression, and even death in severe cases. As a result, there are ongoing global efforts to better understand the genetics, biochemistry, and regulation of aflatoxin biosynthesis, as well as the taxonomy, biology, toxicology, and evolution of aflatoxigenic fungi.

Some of the key areas of research in this field include identifying the genes and pathways involved in aflatoxin biosynthesis [38], developing new methods for detecting and quantifying these toxins, and exploring potential strategies for preventing or reducing aflatoxin contamination in crops. By gaining a better understanding of these factors, researchers hope to develop new tools and approaches for mitigating the risk of aflatoxin exposure and protecting public health.

7. Understanding the genetics and biochemistry of aflatoxin biosynthesis: implications for food safety and public health

Identifying the genes and pathways involved in aflatoxin biosynthesis is an important area of research for several reasons. First, it can help us to better understand the mechanisms by which aflatoxins are produced by fungi, which can in turn inform strategies for preventing or reducing their production. Second, it can help us to develop new tools for detecting and quantifying aflatoxins in crops and food products, which is essential for ensuring food safety. Over the years, researchers have made significant progress in identifying the genes and pathways involved in aflatoxin biosynthesis. This work has been facilitated by advances in genomics and bioinformatics, which have enabled researchers to sequence and analyze the genomes of aflatoxin-producing fungi. Some of the key genes and pathways involved in aflatoxin biosynthesis include the following:

1. The aflatoxin biosynthetic gene cluster (AF cluster): This cluster contains around 25 genes that are involved in the biosynthesis of different intermediates in the aflatoxin pathway [39]. The cluster is regulated by a complex network of

transcription factors that respond to environmental signals, such as temperature [40], pH [41], and nutrient availability [42, 43].

2. Polyketide synthase (PKS) and non-ribosomal peptide synthetase (NRPS) enzymes: These enzymes are responsible for assembling the complex carbon skeletons of the aflatoxin molecules. There are several PKS and NRPS enzymes involved in the biosynthesis of aflatoxins, each of which produces a different intermediate in the pathway [44]. The genes encoding these enzymes are often clustered in fungal genomes and are regulated by complex networks of transcription factors, which can respond to environmental cues such as temperature, pH, nutrient availability, and stress. Additionally, the production of mycotoxins can be influenced by the interactions between fungi and their hosts, as well as by the presence of other microorganisms in the same environment [45].

3. Cytochrome P450 monooxygenases: Cytochrome P450 enzymes are a family of heme-containing enzymes that play a crucial role in the metabolism of xenobiotics, including drugs and environmental pollutants. Cytochrome P450 enzymes are highly variable and have evolved to meet the demands of different organisms and environments. These enzymes are responsible for oxidizing the intermediates produced by the PKS and NRPS enzymes, which creates the different types of aflatoxins [46].

By studying the function of these genes and their products, researchers can gain a deeper understanding of how aflatoxins are produced and regulated in fungi, and develop new approaches for preventing or reducing their production. Payne & Brown [47] highlights the importance of understanding the genetic and physiological mechanisms of aflatoxin biosynthesis in developing effective strategies for controlling its production and minimizing its impact on human and animal health.

8. Advances in methods for detecting and quantifying aflatoxins in crops and food products: implications for food safety

Monitoring and controlling aflatoxin contamination in food products is crucial to ensure food safety and protect public health. By establishing reliable testing methods and implementing prevention strategies, we can minimize the risk of exposure to aflatoxins and mitigate their adverse health effects [48]. Developing new methods for detecting and quantifying aflatoxins is essential for ensuring food safety and protecting public health. Traditional methods for detecting aflatoxins include thin-layer chromatography (TLC) [49] and high-performance liquid chromatography (HPLC), which are effective but time-consuming and labor-intensive [50]. In recent years, several new methods have been developed for detecting and quantifying aflatoxins in crops and food products. Some of these methods include the following:

8.1 Immunoassays

Immunoassays are based on the use of antibodies that specifically recognize and bind to aflatoxins. Immunoassays can be performed using a range of formats [51], including ELISA (enzyme-linked immunosorbent assay) [52, 53], lateral flow tests [54], and fluorescent assays [55].

8.2 Mass spectrometry

Mass spectrometry is a powerful analytical technique that can detect and quantify the presence of aflatoxins with high sensitivity and specificity. Mass spectrometry can be coupled with different separation techniques, such as liquid chromatography (LC) or gas chromatography (GC), to achieve high levels of separation and detection [56].

8.3 Biosensors

Biosensors are devices that use biological components, such as enzymes or antibodies, to detect and quantify the presence of aflatoxins in food samples [57]. Biosensors can be based on different transduction principles, such as electrochemical, optical, or piezoelectric transduction [58].

8.4 DNA-based methods

DNA-based methods use DNA probes or PCR (polymerase chain reaction) to detect the presence of aflatoxin-producing fungi in crops or food products. DNA-based methods can be faster and more sensitive than traditional methods, but they require specialized equipment and expertise [59].

By developing new methods for detecting and quantifying aflatoxins, researchers can improve our ability to monitor and control the presence of these toxins in food products, and ensure that they meet regulatory standards for food safety.

9. Aflatoxin prevention for food safety

Exploring potential strategies for preventing or reducing aflatoxin contamination in crops is critical for ensuring food safety and protecting public health. Some of the key strategies that researchers are investigating in this area include the following:

9.1 Breeding for resistance against aflatoxin contamination

Breeding groundnut varieties with stable resistance to aflatoxin contamination is a sustainable and effective approach to reducing the problem. However, this poses challenges to breeders due to the limited availability of improved germplasm and the significant genotype-by-environment (GxE) interaction for aflatoxin contamination. The limited germplasm restricts the range of genetic variability that breeders can work with, making it difficult to identify suitable parental lines for breeding. Additionally, the significant GxE interaction means that the performance of a genotype in one environment cannot be used to predict its performance in another environment, making it hard to select for resistance across diverse environments. Overcoming these challenges will require breeders to utilize innovative breeding techniques, such as marker-assisted selection, genomic selection, and multi-environment testing, to identify and incorporate favorable alleles for resistance across diverse environments [60, 61]. To combat this problem, researchers have been working on developing crops that are resistant to aflatoxin contamination. Aflatoxin biosynthesis in *Aspergillus* spp. is regulated by oxidative stress responses, which are induced by environmental stresses such as drought and heat stress. Host-derived reactive oxygen species (ROS) may play a role in cross-kingdom communication between host plants and *A. flavus*.

9.1.1 Application of RNA interference (RNAi) technology

Recent advances in plant breeding technology have enabled the study and application of metabolomic, proteomic, and transcriptomic knowledge in productive breeding populations [62]. Researchers are exploring ways to engineer crops that are resistant to aflatoxin-producing fungi or that produce antifungal compounds that can inhibit the growth of these fungi. However, there have been some promising advancements in engineering aflatoxin-resistant crops. One approach is to use RNA interference (RNAi) technology to silence genes in the fungi responsible for aflatoxin production [63, 64]. Another approach is to introduce genes from other organisms, such as bacteria or plants, that can break down aflatoxins [65]. Breeders now have more options for their improvement programs as there are a greater number of maize breeding lines that demonstrate resistance to both *A. flavus* infection and aflatoxin accumulation. The majority of these resistant lines have a tropical background, but newer lines have been created through crosses and backcrosses between tropical and temperate germplasm [66]. The GEM project has been instrumental in developing many of these lines through breeding crosses and hybrids [67]. Genomic regions that offer a consistent increase in resistance to aflatoxin or *A. flavus* induced ear rot in resistant maize lines have been identified through various studies, including several QTL and one meta-QTL analysis. Each of these QTL typically accounts for 5–20% of the observed variation in resistance. However, it is possible that the effects of some of these QTL may have been overestimated due to the Beavis effect [68].

9.1.2 The introduction of genes from other organisms

Another approach is to use genetic engineering to introduce genes that can enhance plant defense mechanisms against fungi, such as the expression of antifungal proteins or enzymes [69]. For example, researchers have successfully introduced genes from a wild peanut species into commercial peanut cultivars to increase their resistance to aflatoxin-producing fungi [70]. Overall, while there are still many challenges to overcome in developing aflatoxin-resistant crops, there is promising progress being made. The development of such crops could have significant benefits in improving food safety, reducing health risks, and increasing food security, especially in developing countries where aflatoxin contamination is a major problem.

10. Crop management practices

Aflatoxin contamination is strongly influenced by environmental factors, such as temperature, humidity, and rainfall, as well as cultural practices, such as planting density, irrigation, and fertilization. Researchers are exploring ways to optimize these factors to reduce the risk of aflatoxin contamination in crops. To prevent the adverse health effects of aflatoxin contamination, it is important to implement appropriate prevention and control measures.

10.1 Preharvest amendments

Preharvest amendments are treatments or additives applied to crops or soil before harvest in order to improve crop quality, yield, or post-harvest performance. The use of preharvest amendments is common in agricultural practices, as they can help to address a range of issues affecting crop production and quality. For example, preharvest amendments may be used to control pests and diseases, improve soil fertility and nutrient availability, enhance plant growth and development, or reduce post-harvest losses and spoilage. However, it is important to carefully consider the potential risks and benefits associated with the use of preharvest amendments, as some may have negative impacts on the environment or human health. In addition, regulations governing the use of preharvest amendments may vary depending on the crop, location, and other factors.

10.1.1 The role of high-quality seeds in preventing aflatoxin contamination in agriculture

Using high-quality seeds is an important part of preventing aflatoxin contamination in food production. High-quality seeds are typically produced using good agricultural practices, such as careful selection and handling of parent plants, testing for disease and pests, and appropriate storage and transport conditions. By using high-quality seeds, farmers can ensure that their crops are healthy and better able to resist fungal infections that can lead to aflatoxin contamination. This is because healthy plants are better able to defend themselves against pathogens and are less likely to develop the kinds of stress conditions that can make them more susceptible to fungal infections.

10.1.2 Managing aflatoxin contamination in groundnuts through soil amendments

Ijaz et al. [71] proposed the use of soil amendments has been as a potential strategy for managing aflatoxin contamination in groundnuts. Several studies have investigated the effects of different soil amendments on aflatoxin levels in groundnuts, including organic amendments such as poultry manure, vermicompost, and green manure, as well as Inorganic amendments such as lime and sulfur. Overall, the results of these studies have been mixed, with some studies reporting significant reductions in aflatoxin levels following the application of soil amendments, while others have found little to no effect. The effectiveness of soil amendments in reducing aflatoxin levels may depend on a variety of factors, including the type and amount of amendment used, the timing and frequency of application, and the local environmental conditions. Despite the inconsistent results, the use of soil amendments may still be a promising approach for managing aflatoxin contamination in groundnuts, particularly in areas where other strategies, such as chemical fungicides, are not available or feasible. However, more research is needed to better understand the mechanisms underlying the effects of soil amendments on aflatoxin contamination, as well as to optimize their use in different agricultural settings.

10.1.3 The importance of crop rotation in preventing aflatoxin contamination

Proper crop rotation is an important practice that can help reduce the risk of aflatoxin contamination in agriculture. This is because crop rotation helps to interrupt the life cycle of the fungi that produce aflatoxins, which can help prevent the buildup of fungal spores and reduce the risk of contamination in subsequent crops. Crop rotation involves planting different crops in a field over successive growing seasons. This helps to break the cycle of plant-specific pests and diseases, which can build up in the soil over time and infect subsequent crops. By alternating crops, farmers can help to disrupt the life cycle of these pests and diseases, which can help to reduce their populations and limit their impact on subsequent crops. In the case of aflatoxin contamination, crop rotation can help to reduce the buildup of fungal spores in the soil. This is because the fungi that produce aflatoxins typically infect specific crops, such as corn and peanuts. By rotating these crops with other types of crops, such as legumes, cereals, or grasses, farmers can help to reduce the buildup of fungal spores in the soil, which can in turn reduce the risk of contamination in subsequent crops [72–74].

10.1.4 Preventing aflatoxin contamination through effective Pest control measures

Aflatoxigenic fungi can grow on crops both in the field and during storage, and their growth is often facilitated by the presence of pests and insects. Pests such as insects and rodents can damage crops, creating entry points for fungi that produce aflatoxins. Proper pest control measures can help prevent crop damage and reduce the risk of fungal infections. Effective pest control measures can include the use of chemical or biological pesticides, crop sanitation practices, and appropriate storage and transport practices. Integrated pest management (IPM) is a holistic approach to pest control that uses a combination of techniques to minimize the use of pesticides while still effectively managing pests. IPM can be an effective approach to preventing aflatoxin contamination while also minimizing the use of potentially harmful chemicals in food production [75].

10.1.5 Biological control of aflatoxins

10.1.5.1 Using non-toxigenic strains of Aspergillus *spp.*

Biological control of aflatoxin using non-toxigenic strains of *Aspergillus* spp. has shown promising results in both laboratory and field studies. These non-toxigenic strains can outcompete and displace the toxigenic strains, reducing the overall levels of aflatoxin contamination in crops [3]. The use of non-toxigenic strains for biological control of aflatoxin is considered to be a safe and environmentally friendly approach. These strains are naturally occurring and do not produce harmful toxins, making them ideal for use in agricultural settings [76]. Several non-toxigenic strains of *Aspergillus* spp. have been identified and tested for their ability to control aflatoxin contamination in crops, including *Aspergillus flavus* AF36 and *Aspergillus parasiticus* NRRL 2999. Studies have shown that these strains can significantly reduce the levels of aflatoxin contamination in crops, including maize, peanuts, and tree nuts. PCR assays can be used to distinguish between genetically similar toxigenic and atoxigenic isolates of *A. flavus* [77]. Specifically, atoxigenic isolates with deletions within the aflatoxin gene cluster can be readily identified using this method. While the use of non-toxigenic strains for biological control of aflatoxin is still in the early stages of development, it shows great potential as a strategy for reducing aflatoxin contamination in crops and improving food safety. Further research is needed to better understand the effectiveness of this approach in different agricultural settings and to develop practical methods for implementing it on a large scale [78].

10.1.5.2 Effectiveness of Trichoderma species and cattle dung as soil amendment in reducing aflatoxin contamination in groundnut

A study conducted recently aimed to investigate the effect of using *Trichoderma* species in combination with cattle dung as a soil amendment on the yield and preharvest aflatoxin contamination of groundnut. The researchers found that the application of *Trichoderma* species, in combination with cattle dung, significantly improved the yield of groundnut and reduced preharvest aflatoxin contamination. Moreover, the study also showed that the application of *Trichoderma* species and cattle dung improved soil fertility and increased the availability of nutrients in the soil. The study suggests that the use of *Trichoderma* species in combination with cattle dung can be an effective strategy to improve crop productivity and reduce aflatoxin contamination in groundnut cultivation [79].

10.2 Post-harvest management

Post-harvest management plays a crucial role in reducing aflatoxin contamination in crops. Aflatoxins are a group of mycotoxins produced by the fungus *Aspergillus flavus* and *Aspergillus parasiticus* that can contaminate crops during preharvest, harvest, and post-harvest stages. Here are some post-harvest management practices that can help reduce aflatoxin contamination:

10.2.1 Mitigating aflatoxin contamination through appropriate storage and drying conditions

Proper storage and drying conditions are critical to mitigating the risk of aflatoxin contamination in food production. After harvest, crops should be stored in dry, cool, and well-ventilated facilities to prevent the growth of fungi that can produce aflatoxins. Crops dried to the appropriate moisture content prevent fungal growth and minimize the risk of contamination. Proper drying conditions may vary depending on the type of crop and the local climate, but generally involve careful monitoring of temperature, humidity, and airflow. Drying can be done using a variety of methods, such as natural drying in the sun or with fans and heaters, or using specialized equipment such as dryers and dehumidifiers. Proper storage and drying practices can significantly reduce the risk of aflatoxin contamination in food production and help ensure the safety and quality of food products [80]. Mousavi Khaneghah et al., [81] emphasizes the importance of understanding and addressing the risks associated with aflatoxin contamination in cereals. While the production of these toxins is a natural occurrence, there are several factors that can contribute to increased contamination, such as improper storage and handling practices.

11. Conclusion

Food security is crucial for human health and socioeconomic stability, but mycotoxins produced by fungi can contaminate crops, leading to health problems and economic losses. Aflatoxin is a mycotoxin that commonly contaminates crops such as corn, peanuts, and cottonseed, and can cause serious health consequences. Preventing and managing aflatoxin contamination requires effective strategies such as improved crop management practices, post-harvest handling and storage, and strict regulation and monitoring of food quality and safety. There are several strategies that researchers are exploring to prevent or reduce aflatoxin contamination in crops. Some of these strategies include the following:

Good agricultural practices: One of the primary ways to prevent aflatoxin contamination is to implement good agricultural practices, such as crop rotation, proper irrigation, and the use of high-quality seeds. These practices can help minimize the growth of *Aspergillus* fungi, which are responsible for producing aflatoxins.

Biological control: Researchers are also investigating the use of biological control agents, such as non-aflatoxigenic strains of *Aspergillus* fungi or other microorganisms, to reduce the growth of aflatoxin-producing fungi. This strategy involves introducing these microorganisms into the soil or onto crops to compete with and displace the aflatoxin-producing strains.

Chemical control: Chemical control methods, such as the use of fungicides or insecticides, can also be effective in preventing aflatoxin contamination. However, these methods can be costly and may have negative environmental impacts.

Post-harvest interventions: Another strategy for reducing aflatoxin contamination is to implement post-harvest interventions, such as proper drying and storage techniques, to prevent the growth of *Aspergillus* fungi and minimize aflatoxin production.

Genetic modification: Researchers are also investigating the use of genetic modification to develop crops that are more resistant to aflatoxin contamination. This strategy involves modifying the genetic makeup of crops to enhance their ability to resist *Aspergillus* fungi and reduce the production of aflatoxins. Breeding groundnut varieties with stable resistance to aflatoxin contamination is a sustainable approach but poses challenges due to limited germplasm and significant genotype-by-environment interaction. Overcoming these challenges will require breeders to utilize innovative breeding techniques and strategies, through collaborative efforts among breeders, geneticists, and agronomists.

In summary, preventing the adverse health effects of aflatoxin contamination requires a multi-pronged approach that involves implementing appropriate prevention and control measures, regular monitoring and testing, and appropriate food safety regulations, as well as promoting public awareness and education about the risks of aflatoxin exposure. In addition to safeguarding the health of consumers, such practices can enhance the quality and market value of their produce. Overall, a combination of these strategies may be necessary to effectively prevent or reduce aflatoxin contamination in crops and ensure food safety.

Conflict of interest

None.

Author details

Amir Afzal[1*], Sairah Syed[1], Rafiq Ahmad[1], Muhammad Zeeshan[2] and Ghulam Nabi[3]

1 Barani Agricultural Research Institute, Chakwal, Pakistan

2 Barani Agricultural Research Station, Fateh Jang, Pakistan

3 Groundnut Research Station, Attock, Pakistan

*Address all correspondence to: rajaamirafzal@gmail.com

References

[1] Vågsholm I, Arzoomand NS, Boqvist S. Food security, safety, and sustainability—Getting the trade-offs right. Front. Sustain. Food Syst. 2020;**4**:16. DOI: 10.3389/fsufs.2020.00016

[2] Kumar A, Pathak H, Bhadauria S, Goyal AK. Aflatoxin contamination in food crops: Causes, detection, and management: A review. Food Prod. Process. Nutr. 2021;**3**(1):17. DOI: 10.1186/s43014-021-00064-y

[3] Udomkun P, Wiredu AN, Nagle M, Müller J, Vanlauwe B, Bandyopadhyay R. Innovative technologies to manage aflatoxins in foods and feeds and the profitability of application–a review. Food Control. 2017;**76**:127-138. DOI: 10.1016/j.foodcont.2017.01.008

[4] Nazhand A, Durazzo A, Lucarini M, Souto EB, Santini AJF. Characteristics, occurrence, detection and detoxification of aflatoxins in foods and feeds. Food. 2020;**9**(5):644

[5] Kumar P, Mahato DK, Kamle M, Mohanta TK, Kang SG. Aflatoxins: A global concern for food safety, human health and their management. Frontiers in Microbiology. 2017;7:2170. DOI: 10.3389/fmicb.2016.02170

[6] Emmott A. Aflatoxins: Finding Solutions for Improved Food Safety. Market-Led Aflatoxin Interventions: Smallholder Groundnut Value Chains in Malawi. IFPRI. (International Food Policy Research Institute) A member of the CGIAR Consortium | A world free of hunger and malnutrition 2033 K Street, NW, Washington, DC 20006-1002 USA; 2013:1-2

[7] Eskola M, Kos G, Elliott CT, Hajšlová J, Mayar S, Krska R. Worldwide contamination of food-crops with mycotoxins: Validity of the widely cited 'FAO estimate' of 25%. Critical Reviews in Food Science and Nutrition. 2020;**60**(15):2773-2789

[8] Ahmad MM, Ahmad M, Ali A, Hamid R, Javed S, Abdin MZ. Detection of *aspergillus flavus* and *aspergillus parasiticus* from aflatoxin-contaminated peanuts and their differentiation using PCR-RFLP. Annales de Microbiologie. 2014;**64**:1597-1605

[9] Gholami-Shabani M, Shams-Ghahfarokhi M, Razzaghi-Abyaneh M. Aflatoxins and aflatoxigenic fungi in Iran: A systematic review of the past, present, and future. Mycol. Iran. 2018;**1**:65-84. DOI: 10.22043/MI.2017.116760

[10] Gourama H, Bullerman LB. *Aspergillus flavus* and *aspergillus parasiticus*: Aflatoxigenic fungi of concern in foods and feeds - a review. Journal of Food Protection. 1995;**58**(12):1395-1404

[11] Campione E, Gaziano R, Doldo E, Marino D, Falconi M, Iacovelli F, et al. Antifungal effect of all-trans retinoic acid against *aspergillus fumigatus* In vitro and in a pulmonary Aspergillosis In vivo model. Antimicrobial Agents and Chemotherapy. 2021;**65**(3):e01874-e01820. DOI: 10.1128/AAC.01874-20

[12] Samson RA. Current taxonomic schemes of genus *aspergillus* and its teleomorphs. Biotechnol. 1992;**23**:355

[13] Frisvad JC, Hubka V, Ezekiel CN, Hong SB, Novakova A, Chen AJ, et al. Taxonomy of *aspergillus* section Flavi and their production of aflatoxins, ochratoxins and other mycotoxins.

Studies in Mycology. 2019;**93**:1-63. DOI: 10.1016/j.simyco.2018.06.001

[14] Probst C, Bandyopadhyay R, Cotty PJ. Diversity of aflatoxin-producing fungi and their impact on food safety in sub-Saharan Africa. International Journal of Food Microbiology. 2014;**174**:113-122. DOI: 10.1016/j.ijfoodmicro.2013.12.010

[15] Cotty PJ. Virulence and cultural characteristics of two *aspergillus flavus* strains pathogenic on cotton. Phytopathology. 1989;**79**:808-814

[16] Cortés G, Carvajal M, Méndez-Ramírez I, Avila-González E, Chilpa-Galván N, Castillo-Urueta P, et al. Identification and quantification of aflatoxins and aflatoxicol from poultry feed and their recovery in poultry litter. Poultry Science. 2010;**89**(5):993-1001. DOI: 10.3382/ps.2009-00019

[17] Bbosa GS, Kitya D, Odda J, Ogwal-Okeng J. Aflatoxins metabolism. Health. 2013;**5**:14-34. DOI: 10.4236/health.2013.57A3003

[18] Vijaya Kumar V. Aflatoxins: Properties, toxicity and detoxification. Nutr. Food Sci. Int. J. 2018;**6**:555696

[19] Magan N, Aldred D. Post-harvest control strategies: Minimizing mycotoxins in the food chain. International Journal of Food Microbiology. 2007;**119**:131-139

[20] Luo S, Du H, Kebede H, Liu Y, Xing F. Contamination status of major mycotoxins in agricultural product and food stuff in Europe. Food Control. 2021;**127**:108120

[21] Abdin M, Ahmad MM, Javed S. Advances in molecular detection of *aspergillus*: An update. Archives of Microbiology. 2010;**192**:409-425

[22] Benkerroum N. Aflatoxins: Producing-molds, structure, health issues and incidence in southeast Asian and sub-Saharan African countries. IJERPH. 2020;**17**:1215. DOI: 10.3390/ijerph17041215

[23] Alshannaq A, Yu JH. Occurrence, toxicity, and analysis of major mycotoxins in food. IJERPH. 2017;**14**(6):632

[24] Santini A, Ritieni A. Aflatoxins: Risk, exposure and remediation. In: Aflatoxins: Recent Advances and Future Prospects. London, UK, London, UK: Intech Open; 2013. pp. 343-376

[25] Coppock RW, Christian RG, Jacobsen BJ. Aflatoxins. In: Gupta RC. editor. Veterinary Toxicology. 3rd ed. London, UK; Academic Press; 2018. pp. 983-994. DOI: 10.1016/B978-0-12-811410-0.00069-6

[26] Kortei NK, Annan T, Dzikunoo J, Mensah DL, Ababio PF. Exposure assessment and risk characterization of aflatoxins intake through consumption of maize (Zea mays) in different age populations in the Volta region of Ghana. Food Contam. 2022;**9**(1):13

[27] Wild CP, Gong YY. Mycotoxins and human disease: A largely ignored global health issue. Carcinogenesis. 2010;**31**(1):71-82

[28] Marr KA, Platt A, Tornheim JA, Zhang SX, Datta K, Cardozo C, et al. Aspergillosis complicating severe coronavirus disease. Emerging Infectious Diseases. 2021;**27**(1):18-25

[29] Salmanton-García J, Sprute R, Stemler J, Bartoletti M, Dupont D, Valerio M, et al. COVID-19-associated pulmonary Aspergillosis, march-august 2020. Emerging Infectious Diseases. 2021;**27**(4):1077-1086

[30] Williams JH, Phillips TD, Jolly PE, Stiles JK, Jolly CM, Aggarwal D. Human aflatoxicosis in developing countries: A review of toxicology, exposure, potential health consequences, and interventions. The American Journal of Clinical Nutrition. 2004;**80**(5):1106-1122. DOI: 10.1093/ajcn/80.5.1106

[31] Kaale L, Kimanya M, Macha I, Mlalila N. Aflatoxin contamination and recommendations to improve its control: A review. World Mycotoxin Journal. 2021;**14**(1):27-40

[32] Dhakal A, Hashmi MF, Sbar E. Aflatoxin Toxicity. Treasure Island (FL): StatPearls; 2022

[33] Abrehame S, Manoj VR, Hailu M, Chen YY, Lin YC, Chen YP. Aflatoxins: Source, detection, Clinical Features and Prevention. Processes. 2023;**11**(1):204. https://doi.org/10.3390/pr11010204

[34] Coulombe RA. Biological action of mycotoxins. Journal of Dairy Science. 1993;**76**(3):880-891

[35] Sherif SO, Salama EE, Abdel-Wahhab MA. Mycotoxins and child health: The need for health risk assessment. International Journal of Hygiene and Environmental Health. 2009;**212**:347-368

[36] McGlynn KA, London WT. Epidemiology and natural history of hepatocellular carcinoma. Best Practice & Research. Clinical Gastroenterology. 2005;**19**:3-23. DOI: 10.1016/j.bpg.2004.10.004

[37] Marchese S, Polo A, Ariano A, Velotto S, Costantini S, Severino L. Aflatoxin B1 and M1: Biological properties and their involvement in cancer development. Toxins. 2018;**10**:214

[38] Yu J. Current understanding on aflatoxin biosynthesis and future perspective in reducing aflatoxin contamination. Toxins. 2012;**4**(11):1024-1057. DOI: 10.3390/toxins4111024

[39] Flaherty JE, Payne GA. Overexpression of aflR leads to upregulation of pathway gene transcription and increased aflatoxin production in *aspergillus flavus*. Applied and Environmental Microbiology. 1997;**63**(10):3995-4000

[40] Gallo A, Solfrizzo M, Epifani F, Panzarini G, Perrone G. Effect of temperature and water activity on gene expression and aflatoxin biosynthesis in *aspergillus flavus* on almond medium. International Journal of Food Microbiology. 2016;**217**:162-169

[41] Luo Z, Ren H, Mousa JJ, Rangel DE, Zhang Y, Bruner SD, et al. The PacC transcription factor regulates secondary metabolite production, stress response, but has only minor effects on virulence in the insect pathogenic fungus Beauveria bassiana. Environmental Microbiology. 2017;**19**:788-802

[42] Calvo AM, Wilson RA, Bok JW, Keller NP. Relationship between secondary metabolism and fungal development. Microbiology and Molecular Biology Reviews. 2002;**66**:447-459

[43] Maggio-Hall LA, Wilson RA, Keller NP. Fundamental contribution of beta-oxidation to polyketide mycotoxin production in planta. Molecular Plant-Microbe Interactions. 2005;**18**:783-793. DOI: 10.1094/MPMI-18-0783

[44] Stachelhaus T, Mootz HD, Marahiel MA. The specificity-conferring code of adenylation domains in non-ribosomal peptide synthetases. Chemistry & Biology. 1999;**6**:493-505

[45] Huffman J, Gerber R, Du L. Recent advancements in the biosynthetic

mechanisms for polyketide-derived mycotoxins. Biopolymers. 2010;**93**(9):764-776. DOI: 10.1002/bip.21483

[46] Gonzalez FJ, Gelboin HV. Human cytochromes P450: Evolution and cDNA-directed expression. Environmental Health Perspectives. 1992;**98**:81-85. DOI: 10.1289/ehp.929881

[47] Payne GA, Brown MP. Genetics and physiology of aflatoxin biosynthesis. Annual Review of Phytopathology. 1998;**36**:329-362

[48] Herzallah SM. Determination of aflatoxins in eggs, milk, meat and meat products using HPLC fluorescent and UV detectors. Food Chemistry. 2009;**114**(3):1141-1146

[49] Braithwaite A, Smith FJ, Stock R. Plane Chromatography. In: Chromatographic Methods. London: Chapman and Hall; 1985:24-84. DOI: 10.1007/978-94-009-4093-2_3

[50] Namjoo M, Salamat F, Rajabli N, Hajihoseeini R, Niknejad F, Kohsar F, et al. Quantitative determination of aflatoxin by high performance liquid chromatography in wheat silos in Golestan Province, north of Iran. Iranian Journal of Public Health. 2016;**45**(7):905-910

[51] Nardo FD, Cavalera S, Baggiani C, Chiarello M, Pazzi M, Anfossi L. Enzyme immunoassay for measuring Aflatoxin B1 in legal cannabis. Toxins. 2020;**12**(4):265. DOI: 10.3390/toxins12040265

[52] Chu FS, Fan TS, Zhang GS, Xu YC, Faust S, Mcmahon PL. Improved enzyme-linked immunosorbent assay for aflatoxin Bj in agricultural commodities. Journal of AOAC International. 1987;**70**(5):854-857

[53] Sun DD, Gu X, Li JG, Yao T, Dong YC. Quality evaluation of five commercial enzyme linked immunosorbent assay kits for detecting aflatoxin b1 in feedstuffs. Asian-Australas. J. Anim. Sci. 2015;**28**(5):691

[54] Cvak B, Warth B, Atehnkeng J, Parich A, Moritz A, Sulyok M, et al. Evaluating the performance of lateral flow devices for Total aflatoxins with special emphasis on their robustness under sub-Saharan conditions. Toxins (Basel). 2021;**13**(11):742. DOI: 10.3390/toxins13110742

[55] Zheng MZ, Richard JL, Binder J. A review of rapid methods for the analysis of mycotoxins. Mycopathologia. 2006;**161**:261-273

[56] Iram W, Anjum T, Iqbal M, Ghaffar A, Abbas M. Mass spectrometric identification and toxicity assessment of degraded products of aflatoxin B1 and B2 by Corymbia citriodora aqueous extracts. Scientific Reports. 2015;**5**:14672. DOI: 10.1038/srep14672

[57] Oliveira IS, da Silva Junior AG, de Andrade CAS, Lima Oliveira MD. Biosensors for early detection of fungi spoilage and toxigenic and mycotoxins in food. Current Opinion in Food Science. 2019;**29**:64-79

[58] Polat EO, Cetin MM, Tabak AF, Bilget Güven E, Uysal BÖ, Arsan T, et al. Transducer Technologies for Biosensors and Their Wearable Applications. Biosensors. 2022;**12**(6):385. DOI: 10.3390/bios12060385

[59] Ortega SF, Siciliano I, Prencipe S, Gullino ML, Spadaro D. Development of PCR, LAMP and qPCR assays for the detection of aflatoxigenic strains of *aspergillus flavus* and *A. parasiticus* in hazelnut. Toxins. 2020;**12**(12):757. DOI: 10.3390/toxins12120757

[60] Nigam SM, Waliyar F, Aruna R, Reddy SV. Breeding peanut for resistance

to aflatoxin contamination at ICRISAT. Peanut Sci. 2009;**36**:42-49

[61] Yu B, Huai D, Huang L, Kang Y, Ren X, Chen Y, et al. Identification of genomic regions and diagnostic markers for resistance to aflatoxin contamination in peanut (*Arachis hypogaea* L.). BMC Genetics. 2019;**20**:32. DOI: 10.1186/s12863-019-0734-z

[62] Fountaina JC, Kheraa P, Yanga L, Nayakb SN, Scullye BT, Lee RD, et al. Resistance to *aspergillus flavus* in maize and peanut: Molecular biology, breeding, environmental stress, and future perspectives. Crop J. 2015;**3**:229-237. DOI: 10.1016/j.cj.2015.02.003

[63] Arias RS, Dang PM, Sobolev VS. RNAi-mediated control of aflatoxins in peanut: Method to analyze mycotoxin production and transgene expression in the peanut/*aspergillus* pathosystem. Journal of Visualized Experiments. 2015;**106**:e53398. DOI: 10.3791/53398

[64] Cheng W, Song X-S, Li H-P, Cao L-H, Sun K, Qiu X-L, et al. Host-induced gene silencing of an essential chitin synthase gene confers durable resistance to Fusarium head blight and seedling blight in wheat. Plant Biotechnology Journal. 2015;**13**:1335-1345

[65] Guan Y, Chen J, Nepovimova E, Long M, Wu W, Kuca K. Aflatoxin detoxification using microorganisms and enzymes. Toxins. 2021;**13**(1):46

[66] Smith JS, Williams WP. Aflatoxin accumulation in a maize Diallel cross containing inbred lines with expired plant variety protection. Agronomy. 2021;**11**:2285. DOI: 10.3390/agronomy11112285

[67] Williams WP, Windham GL. Registration of Mp718 and Mp719 germplasm lines of maize. J. Plant Regist. 2012;**6**(2):200-202

[68] Beavis WD. QTL analyses: Power, precision, and accuracy. In: Patterson AH, editor. Molecular Dissection of Complex Traits. New York, NY, USA: CRC Press; 1998. pp. 145-162

[69] Sher Khan R, Iqbal A, Malak R, Shehryar K, Attia S, Ahmed T, et al. Plant defensins: Types, mechanism of action and prospects of genetic engineering for enhanced disease resistance in plants. 3 Biotech. 2019;**9**(5):192. DOI: 10.1007/s13205-019-1725-5

[70] Arias RS, Sobolev VS, Massa AN, et al. New tools to screen wild peanut species for aflatoxin accumulation and genetic fingerprinting. BMC Plant Biology. 2018;**18**:170. DOI: 10.1186/s12870-018-1355-9

[71] Ijaz M, Afzal A, Ali SR, Nawaz HH, Altaf R, Jannat FT, et al. Aflatoxin: Can be managed by soil amendments in groundnut? Pak. J. Phytopath. 2022;**34**(01):65-70. DOI: 10.33866/phytopathol.034.01.0749

[72] Abraham A, Saka V, Mhango W, Njoroge SMC, Brandenburg R. Effect of crop rotation on aflatoxin contamination in groundnuts. In: Proc. 5th RUFORUM Biennial Reg. Conf. Cape Town, South Africa: RUFORUM Working Document Series No. 14(3); 2016. pp. 173-178

[73] Fandohan P, Hell K, Marasas WF. Food processing to reduce mycotoxins in Africa. In: Leslie JF, Bandyopadhyay R, Visconti A, editors. Mycotoxins: Detection Methods, Management, Public Health and Agricultural Trade. Wallingford, UK: CABI (Centre for Agriculture and Bioscience International); 2008. pp. 47-59

[74] Mutegi CK, Ngugi HK, Hendriks SL, Jones RB. Prevalence and factors associated with aflatoxin contamination of peanuts from western

Kenya. International Journal of Food Microbiology. 2009;**130**:27-34

[75] Nesci A, Baara P, Etcheverry M. Integrated management of insect vectors of *aspergillus flavus* in stored maize, using synthetic antioxidants and natural phytochemicals. Journal of Stored Products Research. 2011;**47**(3):231-237

[76] Zanon MSA, Barros GG, Chulze SN. Non-aflatoxigenic *aspergillus flavus* as potential biocontrol agents to reduce aflatoxin 594 contamination in peanuts harvested in northern Argentina. International Journal of Food Microbiology. 2016;**231**:63-68

[77] Mylroie JE, Ozkan S, Shivaji R, Windham GL, Alpe MN, Williams WP. Identification and quantification of a toxigenic and non-toxigenic *aspergillus flavus* strain in contaminated maize using quantitative real-time PCR. Toxins (Basel). 2016;**8**(1):15

[78] Weaver MA, Abbas HK, Falconer LL, Allen TW, Pringle HC III, Sciumbato GL. Biological control of aflatoxin is effective and economical in Mississippi field trials. Crop Protection. 2015;**69**:52-55

[79] Dania VO, Eze SE. Using Trichoderma species in combination with cattle dung as soil amendment improves yield and reduces pre-harvest aflatoxin contamination in groundnut. Agrivita J. Agric. Sci. 2020;**42**:449-461

[80] Pitt JI, Taniwaki MH, Cole MB. Mycotoxin production in major crops as influenced by growing, harvesting, storage and processing, with emphasis on the achievement of food safety objectives. Food Control. 2013;**32**(1):205-215

[81] Mousavi Khaneghah A, Eş I, Raeisi S, Fakhri Y. Aflatoxins in cereals: State of the art. Journal of Food Safety. 2018;**38**:e12532. DOI: 10.1111/jfs.12532

Chapter 10

Whey Protein Fermentation with *Aspergillus niger*: Source of Antioxidant Peptides

Marcela Patricia Gomez Rojas
and Oscar Marino Mosquera Martinez

Abstract

Aspergillus niger is a filamentous fungus that through its proteolytic activity, as a result of its proteases, hydrolyzes whey proteins into smaller peptides. These peptides are characterized by antioxidant properties due to the presence of specific amino acids, such as histidine, tyrosine, tryptophan, cysteine, and methionine, which have been shown to have antioxidant effects. Considering the above, peptide extracts derived from the fermentation of a lactic serum substrate with *Aspergillus niger* were obtained, which were partially purified by precipitation with $ZnSO_4$/acetone; subsequently, the antioxidant capacity was evaluated by spectrophotometric techniques as 2,2-azinobis-3ethyl benzothiazole-6-sulfonic acid ($ABTS^{\blacksquare+}$), diphenylpicrylhydrazyl ($DPPH^{\blacksquare}$), in 96-well microplates, these analyses showed that these extracts have an antioxidant activity higher than 50%; likewise, the amount of thiol groups (-SH) was determined to be higher than 29 nmol/μL and the superoxide dismutase activity (SOD) with values above 0.010 SOD units/mL. For this reason, it is proposed that they can be studied in the future as substances within a food supplementation or in the therapeutic field.

Keywords: *Aspergillus niger*, proteolytic activity, antioxidant activity, fermentation, whey

1. Introduction

Peptides are low molecular weight protein fragments, consisting of ~2 to 20 amino acid residues [1–3], some of which exhibit physiological effects, beneficial to humans, which is why they are called bioactive peptides [4]. These are obtained mainly by hydrolysis of precursor proteins during fermentation processes with exogenous enzymes [5] from plant, bacterial or fungal sources [6], or through their expression and secretion during metabolic processes [7].

Fungal sources have aroused special interest because they exhibit a number of exogenous enzymes with applicability at pharmacological and industrial level [8]; among the most widely used fungal species is the genus *Aspergillus*, whose species *Aspergillus oryzae* [9], *Aspergillus flavipes* and *Aspergillus niger* [10], when used as an enzymatic source gave way to obtain peptides with antihypertensive [11], antioxidant [12], antidiabetic [13], antimicrobial [9], and antioxidant [14] bioactivities, among others.

Likewise, among the sources of precursor proteins used as substrates by fungal enzymatic sources are various kinds of milk and their derivatives [15], gelatin [11], grains such as lentils [13] and soy [16], and even eggs [17]. It should be noted that of these protein sources, milk is the most consumed worldwide [18] and its proteins have different biological and nutritional properties [19], and make it a source of bioactive peptides, which are released from precursor proteins, such as α-lactalbumin (α-LA), β-lactoglobulin (β-LG), caseins (CN), immunoglobulins (Ig), lactoferrin (LF), peptide-protein fractions, phosphoglycoproteins and minor serum proteins (transferrin and serum albumin) [20], during processes already mentioned such as fermentation, chemical hydrolysis or enzymatic hydrolysis [21].

Products derived from fermentation processes have been relevant in people's diets because their nutritional properties are enhanced, thanks to the fact that microorganisms synthesize vitamins, minerals, and bioactive peptides [22, 23], among others, which are beneficial to human health. This is why the evaluation of fermentation processes such as milk fermentation with *Aspergillus niger* arouses interest.

On the other hand, worldwide interest has increased in topics related to conditions caused by oxidative stress since this is related to the development and onset of various human diseases [24, 25], including atherosclerosis [26], Alzheimer's disease [27] and cancer [28], among others.

To analyze the previous problem, a solution was found by evaluating the antioxidant activity through the DPPH$^{\blacksquare}$, ABTS$^{\blacksquare +}$ methodologies, evaluation of thiol groups, and evaluation of superoxide dismutase activity; all these analyses were performed by spectrophotometric techniques in 96-well plates, in the presence of peptide extracts obtained during the fermentation of lactic serum with *Aspergillus niger*.

2. Methodology

2.1 Obtaining the inoculum

The inoculum of *A. niger* CMPUJH002, provided by the collection of microorganisms of the Universidad Pontifica Javeriana, was obtained by spiking on potato dextrose agar (PDA) at 37°C for 7 days. After its use, it was preserved by the method of spore suspension in glycerol of the laboratory of the biotechnology research group — natural products of the Universidad Tecnologica de Pereira (GB-PN, UTP).

2.2 Lactic fermentation

To obtain the peptide extracts, the methodology proposed by Channe & Shewale [29] was used, where lactic serum (enriched and prepared medium) was prepared and sterilized (**Table 1**), as a substrate for *A. niger*, in a 1 L Erlenmeyer, covered with vinipelt plastic. Fermentation was carried out over a period of 8 days, with 400 mL of substrate and 1.6 cm^2 of inoculum of *A. niger* grown on PDA agar after 7 days of growth at 37°C. The fermentation process was carried out with the following conditions pH 4.5, temperature 30°C, and an agitation of 71.76 ± 1.25 rpm, leaving a headspace of 60%. This assay was assembled in quadruplicate, coding each extract

Component	Concentration
Meat peptone	20 g/L
Starch	20 g/L
Glucose	5 g/L
Powdered milk	10 g/L
CaCl2.2H2O	1 g/L
MgSO4.7H2O	0.3 g/L
FeSO4.7H2O	5 mg/L
MnSO4.1H2O	1.56 mg/L
ZnSO4.7H2O	5.34 mg/L
COCl2.6H2O	2 mg/L

Table 1.
Composition of lactic whey fermentation medium enriched with Aspergillus niger *taken from Channe & Shewale [29].*

with an F indicating fermentation and a letter D followed by a number indicating the day it was collected.

2.3 Partial purification of peptide extracts

The extracts collected daily were added 10% trichloroacetic acid [30], in equal proportion with respect to the sample (1, 1), in order to precipitate the large proteins and eliminate contaminants, then the precipitation of low molecular weight proteins was carried out by adding 50 mM ZnSO4/Acetone 80% to the sample in a ratio of 0.25:0.25:0.5 with respect to the sample, then it was placed in a refrigerator at 2°C for 24 hours for subsequent centrifugation at 5000 rpm at 3°C [31].

2.4 Antioxidant activity against DPPH radicals

A solution of DPPH▪ 20 mg/L in methanol was prepared, and 100 μL of this solution was added to each of the wells containing 25 μL of the extract to be evaluated; it was left to react for 30 minutes in darkness and after this time the absorbance was read at 517 nm [25] using the Thermo Scientific™ Multiskan™ GO Microplate Spectrophotometer. This analysis was carried out in quadruplicate.

2.5 ABTS $^+$ antioxidant activity

A solution of ABTS 3.5 mM and potassium persulfate 1.25 mM dissolved in H2O was prepared (this mixture was made at least 12 hours before its use), after this time the absorbance of the ABTS▪ + solution was adjusted to 0.7+/−0.02 units at 732 nm with ethanol; 194 μL of this solution was transferred to the well containing 6 μL of the extract to be evaluated, and it was left to react for 30 minutes in darkness to read its absorbance at 732 nm, in the Thermo Scientific™ Multiskan™ GO Microplate Spectrophotometer. This analysis was carried out in quadruplicate [32].

From the absorbances obtained, the percentage of antioxidant activity is determined with the following equation described by Perez and coworkers [25, 32].

$$\%\text{Antioxidant Activity} = \left[\frac{A_{control\ (-)} - (A_{extract} - A_{white\ extract})}{A_{control}}\right] x100 \qquad (1)$$

Where,
Aextract: Absorbance of the extracts.
Ablanco extracto: Absorbance of the blank of the extracts.
Acontrol (−): Absorbance of the negative control.

2.6 Content analysis of thiol groups (-SH)

For the quantification of thiols, the reaction of the samples with Ellmann's reagent (DNTB (5,5'-Dithiobis(2-nitrobenzoic acid)) was carried out, taking 95 μL of sample with 30 μL of Na2HPO4 buffer, 0.5 M pH 7.0, with 125 μL DNTB at 10 nM incubating for 15 minutes and reading their absorbances at 412 nm, on Thermo Scientific™ Multiskan™ GO Microplate Spectrophotometer equipment; the calibration curve was prepared at different concentrations using glutathione enzyme, Na2HPO4 buffer and sulfoanilic acid 20%(m/v), 125 μL of each standard was taken and analyzing as samples [33].

2.7 Quantification of the enzyme superoxide dismutase (SOD)

The method used for the determination of superoxide dismutase was based on the protocol carried out by Betancur and Mosquera [33], where a hydroxylamine calibration curve is constructed that interacts with a xanthine/xanthine oxidase system as a source of generation of a superoxide anion flux that oxidizes hydroxylamine to nitrite for a subsequent measurement of nitrite concentration by UV/visible spectrometry.

To perform the analysis, sample preparation was done with the addition of 150 μL KH2PO4 buffer, pH 7.8, 40 μL of deionized water, 15 μL of xanthine, 15 μL hydroxylamine hydrochloride 1 mM, and 75 μL xanthine oxidase (0.2 mg protein/mL); the standards for the curve were prepared with phosphate buffer, water, xanthine, xanthine oxidase, hydroxylamine chloride, and the enzyme (SOD), and the system was left to react for 20 minutes in the dark, Then the reaction was stopped in an ice bath to add 100 μL of 19 mM sulfanilic acid and 7 mM α-naphthylamine, after which 100 μL of sample or standard was added, incubated for 20 minutes and its absorbance was measured at 529 nm in the Thermo Scientific™ Multiskan™ GO Microplate Spectrophotometer.

2.8 Protein profile of peptide extracts by denaturing electrophoresis

Electrophoresis was carried out according to Bio-Rad guidelines under denaturing conditions. Separation gels were 15% SDS-polyacrylamide (30% bis-acrylamide, 1.5 M Tris-HCl pH 8.8, 10% ammonium persulfate, 0.04% tetramethylethylenediamine (TEMED), 10% SDS, and distilled water to final volume) and 5% SDS-polyacrylamide stacking gels (30% bis-acrylamide, 1 M Tris-HCl pH 6.8, 10% ammonium persulfate, 0.1% TEMED, 10% sodium dodecyl sulfate (SDS), and distilled water. The solutions were cast in mini protean (Bio-Rad) with the addition of ammonium persulfate and TEMED to polymerize the gels (1 mm thick).

The gels were loaded with 5% concentration samples, which were diluted in 1:1 reducing buffer (Tris HCl (63 mM), glycerol (10%), SDS (2%), bromophenol blue (0.0025%) pH 6.8), heated at 85°C for 2 to 5 minutes; the electrophoresis was performed at room temperature (24 and 27°C) using a constant current of 35 mA at 120 V for approximately 5 h. Once the run was completed, the gels were washed two times with deionized water and stained with a 0.1% Coomassie brilliant blue R-250 solution with 40% methanol and 10% acetic acid for 12 h with gentle agitation and then destained 2 times for 20 minutes with a 25% ethanol and 8% acetic acid solution [34].

2.9 Statistical analysis

For the statistical treatment of the data obtained for antioxidant and antibacterial activity, a one-way ANOVA analysis by replicates with a confidence level of 95% was carried out. All statistical analysis was carried out using GraphPad Prism 8.4.3 software.

3. Results and discussion

3.1 Antioxidant evaluation by means of the DPPH-radical

The DPPH- decolorization methodology carried out allowed establishing that the extracts evaluated have an antioxidant capacity that inhibits the DPPH radical by more than 50%; however, these values are lower than the percentage of antioxidant capacity of the hydroquinone positive control (**Figure 1**). Likewise, these results are lower, compared to other similar works, where the antioxidant activity on DPPH-presented by peptides derived from proteases of *Aspergillus oryzae* and *Aspergillus flavipes* species in milk, exceeded ~90% [9], through the same method, since the method of obtaining the peptides was carried out from the already purified fungal proteases, which leads to their obtaining in a targeted manner, without competition from other enzymes and without the generation of secondary reactions due to the use of the complete metabolism of the microorganism [35].

The results are attributed to the activity coming from the phenolic compounds, which are mainly responsible for the antioxidant activity, being one of the reasons why it is not a recommended technique for samples of biological origin [36], which is why the evaluation of ABTS− + cation radical decolorization was carried out as a complementary technique to contrast the results.

3.2 ABTS^{+} antioxidant evaluation

In this method, the evaluation of antioxidant activity exceeds 90% in almost all cases and can be considered a good antioxidant because the activity significantly exceeds the positive control with equal concentration (1000 ppm) (**Figure 2**).

Although oxidative stress is generated when the oxide-reduction homeostasis state of the cell becomes unbalanced because the antioxidant and prooxidant counterparts are altered. Aerobic organisms activate their defense mechanisms such as the secretion and action of glutathione, which is a tripeptide, composed of glutamate, cysteine, and glycine, that is used at the biological level for the regulation of this oxidation, which is why it is

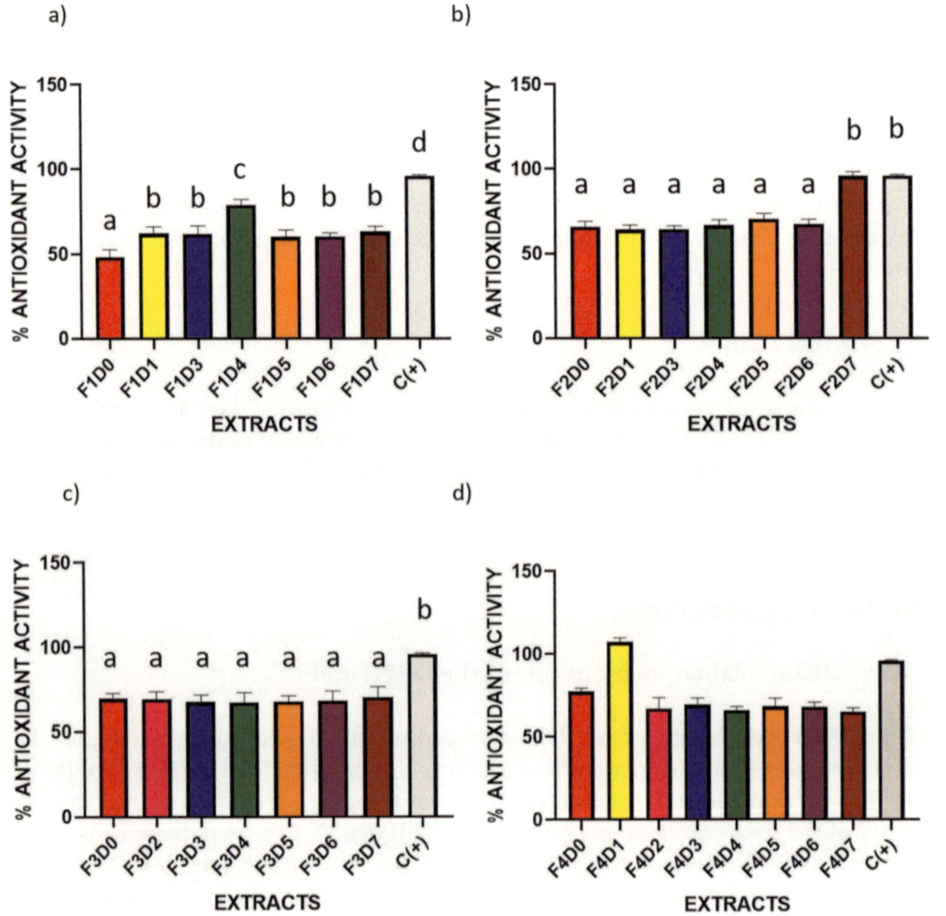

Figure 1.
One-way ANOVA of the determination of antioxidant activity (%AA) by DPPH of fermentations (F) 1, 2, 3, and 4 with their respective days (D) 1, 2, 3, 4, 5, 6, and 7. A) Fermentation 1, b) fermentation 2, c) fermentation 3, and d) fermentation 4. Equal symbol (a; b; c) indicates that there are no significant differences, and equal sign indicates that there are significant differences (P < 0.05).

considered the universal antioxidant, in fact, it has been attributed that the proper functioning of most other antioxidants is due to the presence of glutathione [37].

Therefore, in addition to the biological activities proposed, the determination of thiols (-SH) in the different peptide extracts was also carried out through the construction of a calibration curve as shown in **Figure 3**.

This analysis provided the plot eq. $Y = 0.03769X + 0.02248$, with an r^2 of 0.9927 and a $p < 0.05$, to determine the concentration of glutathione conferring antioxidant activity to the peptides present in each sample, as synthesized in **Table 2**.

To complement the results obtained on the antioxidant potential of peptide extracts, the evaluation of superoxide dismutases (SOD), whose main function is the defense of aerobic organisms such as yeasts and filamentous fungi, such as *Aspergillus oryzae*, was carried out, where through transcriptomic analysis the expression of enzymes, such as catalase, glutathione peroxidase, and superoxide dismutase, has been demonstrated to regulate the concentration of oxidizing agents, such as oxygen free radicals, especially superoxide anion radicals [9]. The results obtained from the

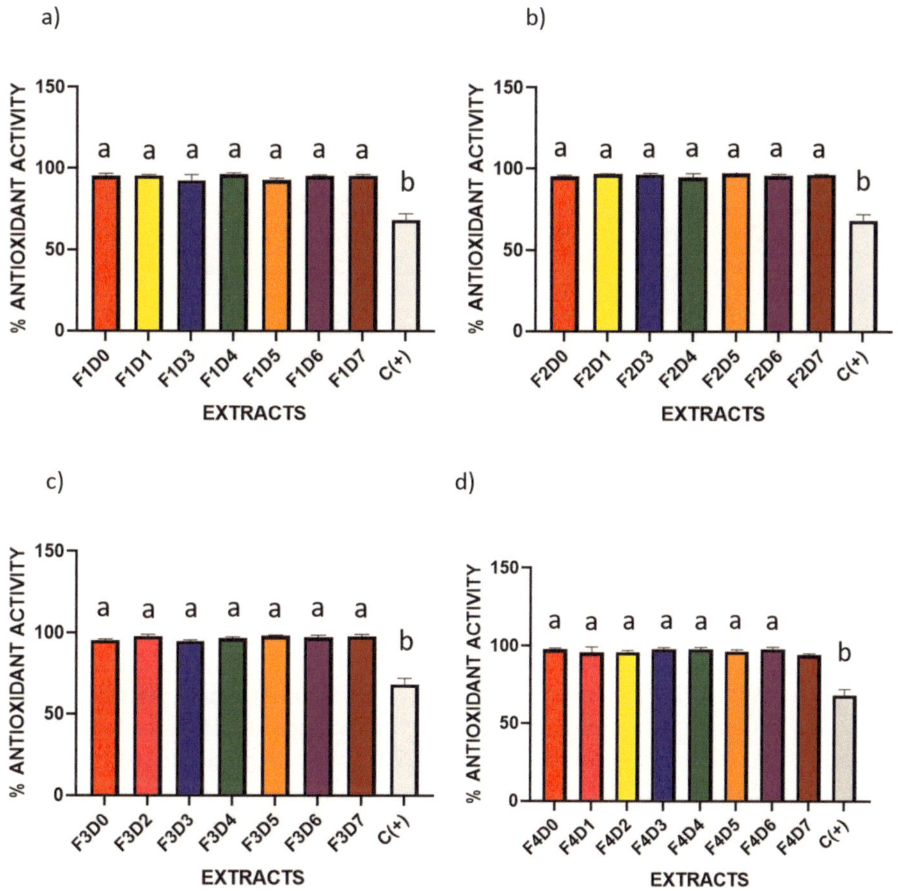

Figure 2.
One-way ANOVA ABTS$^+$ *of fermentations (F) 1, 2, 3, and 4 with their respective days (D) 1, 2, 3, 4, 5, 6, and 7. A) Fermentation 1, b) fermentation 2, c) fermentation 3, and d) fermentation 4. Equal symbol (a; b; c) indicates no difference means, and equal sign indicates significant differences ($p < 0.05$).*

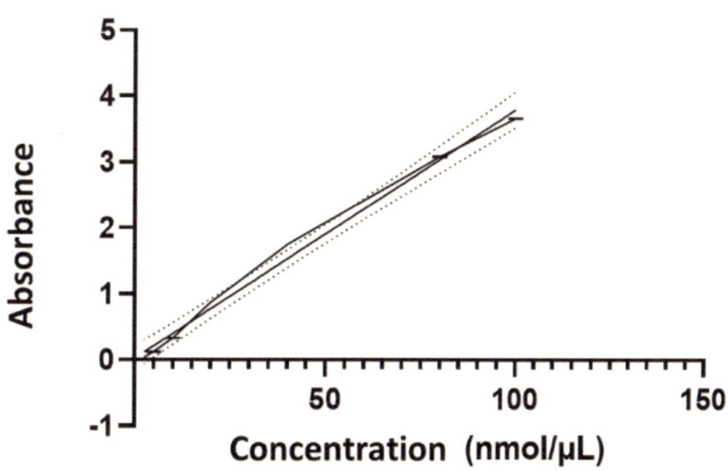

Figure 3.
Calibration curve for thiol (-SH) quantification.

	Concentration (nmol/µL)			
Día	F1	F2	F3	F4
0	39.877 ± 0.109	38.763 ± 0.297	38/763 ± 0.074	38.505 ± 0.126
1	30.983 ± 0.040	38.0571 ± 0.082	-a	41.432 ± 0.068
2	- a	-a	30.382 ± 0.051	33.705 ± 0.135
3	32.749 ± 0.053	30.382 ± 0.022	31.294 ± 0.091	30.375 ± 0.096
4	33.576 ± 0.020	31.294 ± 0.055	30.880 ± 0.063	32.475 ± 0.152
5	35.919 ± 0.133	30.880 ± 0.0178	31.033 ± 0.044	32.695 ± 0.098
6	31.034 ± 0.059	31.033 ± 0.047	29.549 ± 0.049	32.056 ± 0.0027
7	30.870 ± 0.054	29.549 ± 0.056	38.763 ± 0.126	32.308 ± 0.077

- a Missing data are the result of sample loss in storage or low extraction yield.

Table 2.
SH concentration (mmol/µL) of the different peptide extracts.

radical uptake assessment using the superoxide dismutase model are given in **Figure 4**.

It has been identified that the antioxidant capacity of milk and its derivatives are mainly due to sulfur-rich amino acids such as tyrosine and cysteine, vitamins A and E, carotenoids, and enzyme systems such as the enzyme superoxide dismutase (SOD), which are useful so that superoxide radicals (O2–), hydroxyl radicals, and peroxide radicals can be inhibited [9].

This analysis provided the equation of the graph Y = 0.3146X - 0.2602 with an r^2 equal to 0.9933 and a $p < 0.05$; to determine the concentration of SOD units present in each sample, as synthesized in **Table 3**, which has the corresponding calculations of the transformation from -log (SOD units) to SOD units.

This analysis revealed the presence of peptides with SOD-type antioxidant character; although a difference in this concentration is evident in the different fermentations, the reason for these differences cannot be discerned without other analyses,

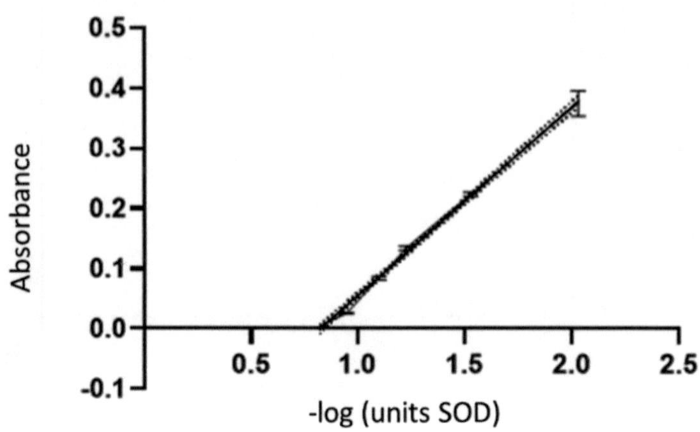

Figure 4.
Calibration curve for SOD determination, –log (SOD units) vs. absorbance.

Concentration (unidades SOD/mL)				
Day	**F1**	**F2**	**F3**	**F4**
0	0.020 ± 0.010	0.030 ± 0.01	0.032 ± 0.005	0.024 ± 0.005
1	0.020 ± 0.020	0.023 ± 0.004	-a	0.035 ± 0.013
2	-a	-a	0.022 ± 0.016	0.022 ± 0.002
3	0.021 ± 0.003	0.021 ± 0.004	0.027 ± 0.004	0.028 ± 0.017
4	0.017 ± 0.007	0.010 ± 0.005	0.010 ± 0.039	0.027 ± 0.011
5	0.019 ± 0.006	0.010 ± 0.004	0.010 ± 0.016	0.032 ± 0.017
6	0.026 ± 0.008	0.010 ± 0.011	0.013 ± 0.02	0.038 ± 0.003
7	0.078 ± 0.004	0.042 ± 0.003	0.011 ± 0.005	0.027 ± 0.014

-a Missing data are the result of sample loss in storage or low extraction yield.

Table 3.
Concentrations in SOD units of the different peptide extracts.

which is why the analysis support is required as a characterization of the peptide extracts.

However, taking into account reports by other authors on the antioxidant activity of raw bovine milk with respect to SOD concentration of 0.92 to 3 units/mL and low concentrations of -SH [38, 39], a decrease is evidenced by the elimination of proteins of higher molecular weight removed during partial purification, that is, concentrated proteins of low molecular weight are good antioxidant agents.

3.3 Analysis of the peptide profile of the extracts by SDS-page electrophoresis

Figure 5 shows the peptide profile exhibited by the peptides obtained, where several fragments of molecular weight above 10 KDa are evident, which led to think that there are still precursor proteins within the extracts; these proteins have already been reported in other studies [40] with their molecular weights as expressed in **Table 4**; therefore, the purification process was not effective in eliminating these high molecular weight proteins.

Likewise, although the technique used does not allow the characterization of peptides of molecular weights lower than 10 KDa, it was the only one that allowed the

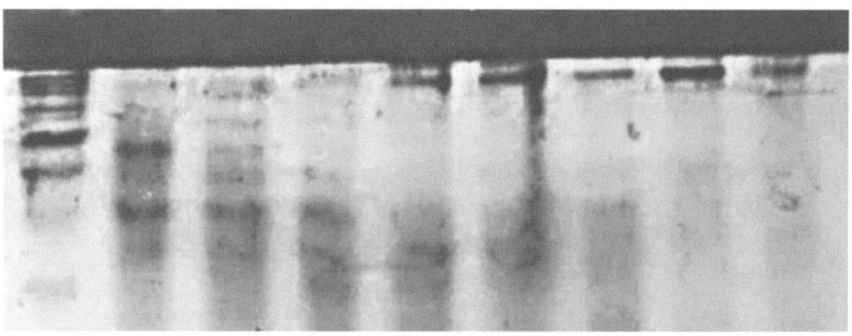

Figure 5.
Electrophoresis, Tris-glycine-SDS for polyacrylamide gel electrophoresis.

Protein	Molecular weight (KDA)	Reported activities	References
Lactoferrin	±80–76	Antiviral, antibacterianas, antivirales, antifúngicas, antiinflamatorias, antioxidantes e inmunomoduladoras	[41–44]
BSA	±66	Antioxidante, antitumoral e inhibidora de la enzima convertidora de angiotensina (ECA)	[45]
Caseins*	±19-30	Antioxidante y antimicrobiana	[46]
β- Lactoglubulin	±18	Antioxidante	[47]
α- Lactoalbumina	±14	Antimicrobiana	[48]

**The hydrolyzates of these proteins exhibit the following activity.*

Table 4.
Molecular weights in KDa of some important proteins in bovine milk.

visualization of bands of protein origin because techniques, such as SDS-triscin [7], did not allow the visualization of bands, as well as silver staining, the result of the electrophoresis can be seen in **Figure 5**; although theoretically it is known that the protein concentration in a sample should be greater than 0.5 mg/mL [34], the samples analyzed have a variable composition and of different concentrations.

4. Conclusions

Antioxidant bioactive peptides have become a very valuable tool in both the food and pharmacological industries due to their ability to act as antioxidants, and thus combat oxidative stress in the human body. These peptides can be obtained from proteins by fermentation processes with microorganisms and enzymes.

In this case, *Aspergillus niger* was used as an enzymatic source to obtain antioxidant peptides from lactic serum. The results obtained indicate that the extracts obtained from this fermentation process have a promising antioxidant activity, which makes them a very interesting nutraceutical substrate.

Importantly, the antioxidant activity of these extracts can be attributed to the presence of thiol groups and positive SOD activity. The ability to catalyze the dismutation of superoxide anion into hydrogen peroxide and molecular oxygen is a very important characteristic of these extracts as it confers them a greater capacity to combat oxidative stress in the human body.

Therefore, the results obtained indicate that the use of *Aspergillus niger* as an enzymatic source for obtaining antioxidant peptides from lactic serum is a very valuable tool in the food and pharmacological industry. These extracts have promising antioxidant activity and are a very interesting source of nutrients to combat oxidative stress in the human body.

Acknowledgements

To the Vice-Rectory of Research of the Universidad Tecnológica de Pereira, for financing project E9-20-3 and to the Universidad de Manizales, where the characterization of the extracts by electrophoresis was carried out.

Conflict of interest

The authors declare no conflict of interest.

Author details

Marcela Patricia Gomez Rojas and Oscar Marino Mosquera Martinez*
Universidad Tecnológica de Pereira, Pereira, Colombia

*Address all correspondence to: omosquer@utp.edu.co

References

[1] Mora L, Aristoy MC, Toldrá F. Bioactive peptides. In: Melton L, Shahidi F, Varelis P, editors. Encyclopedia of Food Chemistry. Amsterdam: Elsevier; 2018. pp. 381-389. DOI: 10.1016/b978-0-08-100596-5.22397-4

[2] Martinez-Villaluenga C, Peñas E, Frias J. Bioactive peptides in fermented foods: Production and evidence for health effects. In: Frias J, Martinez-Villaluenga C, Peñas E, editors. Fermented Foods in Health and Disease Prevention. Amsterdam: Elsevier; 2017. pp. 23-47. DOI: 10.1016/B978-0-12-802309-9.00002-9

[3] Lee JK, Jeon JK, Kim SK, Byun HG. Characterization of bioactive peptides obtained from marine invertebrates. In: Kim S-K, editor. Advances in Food and Nutrition Research. Vol. 65. Amsterdam: Elsevier; 2012. pp. 47-72. DOI: 10.1016/B978-0-12-416003-3.00004-4

[4] Chelliah R et al. The role of bioactive peptides in diabetes and obesity. Food. 2021;**10**(9):1-24. DOI: 10.3390/FOODS10092220/S1

[5] Tüysüz B, Dertli E, Çakir Ö, Şahin E. Bioactive peptides : Formation and impact mechanisms 3 rd international conference on advanced engineering technologies. 3rd International Conference on Advanced Engineering Technolology. 2019;**September**:736-743

[6] Mustafa K, Kanwal J, Musaddiq S, Khakwani S. Bioactive peptides and their natural sources. Functional Foods and Nutraceuticals. 2020; **August**:75-97. DOI: 10.1007/978-3-030-42319-3_5

[7] Jarvis DL, Summers MD, Garcia A, Bohlmeyer DA. Influence of different signal peptides and prosequences on expression and secretion of human tissue plasminogen activator in the baculovirus system. The Journal of Biological Chemistry. 1993;**268**(22):16754-16762. DOI: 10.1016/S0021-9258(19)85481-9

[8] Youssef FS, Ashour ML, Singab ANB, Wink M. A comprehensive review of bioactive peptides from marine fungi and their biological significance. Marine Drugs. 2019;**17**(10):1-24. DOI: 10.3390/MD17100559

[9] Zanutto-Elgui MR et al. Production of milk peptides with antimicrobial and antioxidant properties through fungal proteases. Food Chemistry. 2019;**278**: 823-831. DOI: 10.1016/J.FOODCHEM.2018.11.119

[10] Zhang H, Tang Y, Ruan C, Bai X. Bioactive secondary metabolites from the endophytic aspergillus genus. Records of Natural Products. 2015;**10**(1): 1-16. ISSN: 1307-6167

[11] O'Keeffe MB, Norris R, Alashi MA, Aluko RE, FitzGerald RJ. Peptide identification in a porcine gelatin prolyl endoproteinase hydrolysate with angiotensin converting enzyme (ACE) inhibitory and hypotensive activity. Journal of Functional Foods. 2017;**34**: 77-88. DOI: 10.1016/J.JFF.2017.04.018

[12] Starzyńska-Janiszewska A, Stodolak B, Gómez-Caravaca AM, Mickowska B, Martin-Garcia B, Byczyński Ł. Mould starter selection for extended solid-state fermentation of quinoa. LWT. 2019;**99**:231-237. DOI: 10.1016/J.LWT.2018.09.055

[13] Alves Magro AE, de Castro RJS. Effects of solid-state fermentation and extraction solvents on the antioxidant properties of lentils. Biocatalysis and

Agricultural Biotechnology. 2020; **28**:101753. DOI: 10.1016/J. BCAB.2020.101753

[14] de Castro RJS, Sato HH. A response surface approach on optimization of hydrolysis parameters for the production of egg white protein hydrolysates with antioxidant activities. Biocatalysis and Agricultural Biotechnology. 2015;**4**(1): 55-62. DOI: 10.1016/J.BCAB.2014.07.001

[15] Piovesana S et al. Peptidome characterization and bioactivity analysis of donkey milk. Journal of Proteomics. Apr 2015;**119**:21-29. DOI: 10.1016/J. JPROT.2015.01.020

[16] Chatterjee C, Gleddie S, Xiao CW. Soybean bioactive peptides and their functional properties. Nutrients. 2018;**10** (9):1-16. DOI: 10.3390/NU10091211

[17] Ge H et al. Egg white peptides ameliorate dextran sulfate sodium-induced acute colitis symptoms by inhibiting the production of pro-inflammatory cytokines and modulation of gut microbiota composition. Food Chemistry. 2021;**360**:1-11. DOI: 10.1016/ J.FOODCHEM.2021.129981

[18] Kumari A, Solanki H, Aparna Sudhakaran V. Novel milk and milk products: Consumer perceptions. In: Dairy Processing: Advanced Research to Applications. Singapore: Springer; 2020. pp. 283-299. ISBN: 978-981-15-2607-7

[19] Goulding DA, Fox PF, O'Mahony JA. Milk proteins: An overview. In: Milk Proteins: From Expression to Food. 3rd ed. London, UK: Academic Press; 2020. pp. 21-98. DOI: 10.1016/B978-0-12-815251-5.00002-5

[20] Berry S et al. Defining the Origin and Function of Bovine Milk Proteins through Genomics: The Biological Implications of Manipulation and Modification. 3rd ed. London, UK: Academic Press; 2019. DOI: 10.1016/ B978-0-12-815251-5.00004-9

[21] de Torres Llanez MJ, Cordoba BV, Córdova AFG. Péptidos bioactivos derivados de las proteínas de la leche. Archivos Latinoamericanos de Nutrición. 2005;**55**(2):111-117 Accessed: Jan. 15, 2022. ISSN: 0004-0622

[22] Şanlier N, Gökcen BB, Sezgin AC. Health benefits of fermented foods. Critical Reviews in Food Science and Nutrition. 2019;**59**(3):506-527. DOI: 10.1080/10408398.2017.1383355

[23] Verni M, Rizzello CG, Coda R. Fermentation biotechnology applied to cereal industry by-products: Nutritional and functional insights. Frontiers in Nutrition. 2019;**6**:42. DOI: 10.3389/ FNUT.2019.00042/BIBTEX

[24] Liguori I et al. Oxidative stress, aging, and diseases. Clinical Interventions in Aging. 2018;**13**:757-772. DOI: 10.2147/CIA.S158513

[25] Perez-Jaramillo CC, Sanchez-Peralta WF, Murillo-Arango W, Mendez-Arteaga JJ. Acción antioxidante conjunta de extractos etanólicos de Mollinedia lanceolata, Croton leptostachyus y Siparuna sessiliflora. La Revista de la Academia Colombiana de Ciencias Exactas, Físicas y Naturales. 2017; **41**(158):64-70. DOI: 10.18257/ RACCEFYN.425

[26] Kattoor AJ, Pothineni NVK, Palagiri D, Mehta JL. Oxidative stress in atherosclerosis. Current Atherosclerosis Reports. 2017;**19**(11):1-11. DOI: 10.1007/ S11883-017-0678-6

[27] Tönnies E, Trushina E. Oxidative stress, synaptic dysfunction, and Alzheimer's disease. Journal of

Alzheimer's Disease. 2017;**57**(4): 1105-1121. DOI: 10.3233/JAD-161088

[28] Hayes JD, Dinkova-Kostova AT, Tew KD. Oxidative stress in cancer. Cancer Cell. 2020;**38**(2):167-197. DOI: 10.1016/J.CCELL.2020.06.001

[29] Channe PS, Shewale JG. Influence of culture conditions on the formation of milk-clotting protease by aspergillus Niger MC4. World Journal of Microbiology and Biotechnology. 1997; **14**(1):11-15. DOI: 10.1023/A: 1008808029745

[30] Koontz L. TCA Precipitation. Methods in Enzymology. 2014;**541**:3-10. DOI: 10.1016/B978-0-12-420119-4.00001-X

[31] Baghalabadi V, Doucette AA. Mass spectrometry profiling of low molecular weight proteins and peptides isolated by acetone precipitation. Analytica Chimica Acta. 2020;**1138**:38-48. DOI: 10.1016/J. ACA.2020.08.057

[32] Melo-Guerrero MC, Ortiz-Jurado DE, Hurtado-Benavides AM. Comparison of the composition and antioxidant activity of the chamomile essential oil (Matricaria chamomilla L.) obtained by supercritical fluids extraction and other green techniques. La Revista de la Academia Colombiana de Ciencias Exactas, Físicas y Naturales. 2020;**44**(172):845-856. DOI: 10.18257/ RACCEFYN.862

[33] Betancur YL, Mosquera OM. Cuantificación de tioles libres y superoxido dismutasa (SOD) en extractos metanolicos de plantas de las familias Asteraceae, Euphorbiaceae y Piperaceae. Revista de la Facultad de Ciencias Básicas. 2017;**13**(2):117-122. DOI: 10.18359/RFCB.2748

[34] BioRad. A Guide to Polyacrylamide Gel Electrophoresis and Detection BEGIN. Bogotá, Colombia: Universidad Nueva Granada; 2020. p. 92

[35] Faber K. Biotransformations in Organic Chemistry. 7th ed. California, U.S.: Bio-Rad; 2018. ISBN: 978-3-319-61590-5

[36] Fernandez-Pachon MS, Villaño D, Troncoso AM, García-Parrilla MC. Revisión de los métodos de evaluación de la actividad antioxidante in vitro del vino y valoración de sus efectos in vivo. Archivos Latinoamericanos de Nutrición. 2006;**56**(2):110-122. ISSN: 0004-0622

[37] Corrales LC, Muñoz Ariza MM. Estrés oxidativo: origen, evolución y consecuencias de la toxicidad del oxígeno. Nova. 2012;**10**(18):213-225. ISSN: 1794-2470

[38] Trajchev M, Nakov D, Gjorgoski I. Optimization of enzymatic assay of superoxid dismutase and glutation peroxidase activity in milk whey | repository of UKIM. In: The 2nd International Symposium for Agriculture and Food, 7–9 October, 2015. Ohrid, Republic of Macedonia: Faculty of Agricultural Science and Food; 2015. pp. 43-51 UDC: 637.142.2

[39] Chen J, Lindmark-Månsson H, Åkesson B. Optimisation of a coupled enzymatic assay of glutathione peroxidase activity in bovine milk and whey. International Dairy Journal. 2000; **10**(5–6):347-351. DOI: 10.1016/ S0958-6946(00)00057-1

[40] Costa FF et al. Microfluidic chip electrophoresis investigation of major milk proteins: Study of buffer effects and quantitative approaching. Analytical Methods. 2014;**6**(6):1666-1673. DOI: 10.1039/C3AY41706A

[41] Conneely OM. Antiinflammatory activities of lactoferrin. Journal of the

American College of Nutrition. 2001;**20** (5 Suppl):389S-395S. DOI: 10.1080/07315724.2001.10719173

[42] Van der Strate BWA, Beljaars L, Molema G, Harmsen MC, Meijer DKF. Antiviral activities of lactoferrin. Antiviral Research. 2001;**52**(3):225-239. DOI: 10.1016/S0166-3542(01)00195-4

[43] Orsi N. The antimicrobial activity of lactoferrin: Current status and perspectives. Biometals. 2004;**17**(3): 189-196. DOI: 10.1023/B:BIOM.0000027691.86757.E2

[44] Shea EC, Whitehouse NL, Erickson PS. Effects of colostrum replacer supplemented with lactoferrin on the blood plasma immunoglobulin G concentration and intestinal absorption of xylose in the neonatal calf. Journal of Animal Science. 2009;**87**(6):2047-2054. DOI: 10.2527/jas.2008-1225

[45] Masuelli MA. Bovine Serum Albumin : Properties and Applications. 1st ed. Germany: Springer Science +Business Media; 2020. ISBN: 978-1-53616-787-0

[46] López-Expósito I, Miralles B, Amigo L, Hernández-Ledesma B. Health effects of cheese components with a focus on bioactive peptides. In: Fermented Foods in Health and Disease Prevention. Illinois, U.S.: American Society of Animal Science; 2017. pp. 239-273. DOI: 10.1016/B978-0-12-802309-9.00011-X

[47] Ma S, Yang X, Wang C, Guo M. Effect of ultrasound treatment on antioxidant activity and structure of β-Lactoglobulin using the box-Behnken design effect of ultrasound treatment on antioxidant activity and structure of β-Lactoglobulin using the box-Behnken design. CyTA-Journal Food. 2018;**16**(1):596-606. DOI: 10.1080/19476337.2018.1441909

[48] Permyakov EA, Berliner LJ. α-Lactalbumin: Structure and function. FEBS Letters. 2000;**473**(3):269-274. DOI: 10.1016/S0014-5793(00)01546-5